Driving Data Projects is the invaluable resource I wish I had when beginning my career. This comprehensive guide outlines the essential steps and roles crucial for executing a successful data program. The clarity of its explanations, alongside illustrative charts, makes it a must-read for any data professional.

Kaitlyn Halamuda *CDMP, Senior Manager, Master Data Management, Salesforce*

Making sense of data is a skill required of all leaders today. This book provides a toolkit that can help leaders understand how to better connect data to decision-making.

Joseph Taylor *PhD, Chair, Department of Information Systems and Business Analytics, California State University*

Christine has a long history of successfully implementing data strategies in all size of organisation. In this book, she condensed her tried and tested processes into an accessible, replicable and readable book.

Andy Cotgreave, *Senior Data Evangelist, Tableau*

Driving Data Projects is a roadmap that guides you through the intricacies of data-driven culture, ensuring that your organisation not only survives but thrives in the data-driven era. Highly recommended for those who seek to lead with data, innovate with purpose and transform their organisations.

Dileepa Prabhakar *MBA, Senior Manager of Engineering, T-Mobile*

Driving Data Projects is a comprehensive guide that provides a practical methodology, covering the full cycle of required activities to deliver projects centred on or including, a data element. The work rightly emphasises that data lies alongside people, processes and technology when managing projects and provides practical tools to underpin the related professional practice.

John Burns, *Information LL.M CEng MBCS, Security Risk Analyst*

Data is alive and with their dynamic nature have great potential to lead us to really informed decisions. This book is a comprehensive roadmap from inception to execution of data-driven initiatives.

Professor Raimondo Fanale, *R&D Manager, Intuisco Ltd*

Driving Data Projects delivers a must-read handbook for every business and IT leader trying to create a data-driven culture across their organisation. By skilfully taking the topic beyond the realm of IT, this book provides a practical perspective of the human side of data, transforming it from a technology initiative into a strategic business priority.

Pedro Arellano, *Technology CEO and Founder, Data and Analytics Leader*

Christine's book, long overdue, expertly bridges the gap from learning data best practices to implementing them in organisations. It explains data work as both art and science, providing practical insights to shift from IT-driven projects to enterprise imperatives. The book emphasizes 'data as a service', crucial for organisational success. Beyond data projects, it delves into overall project excellence, from visioning to closure, stakeholder involvement, resistance management and constructive acknowledgement.

Kelle O'Neal, *CEO and Founder, First San Francisco Partners*

So much discussion these days on data culture, strategy, and leadership. Important stuff. But let's not forget to actually do the work, which is where Haskell's *Driving Data Projects* comes in.

Thomas C. Redman PhD, *'the Data Doc', Data Quality Solutions*

Practical, useful, realistic. If you want straight talk on what is needed to have a successful data project, go no further. But don't let the title fool you – every IT project includes data and most overlook or minimise the data aspects. What is offered here should be part of every IT project. Ignore it at your own risk.

Danette McGilvray, *President and Principal, Granite Falls Consulting Inc*

Driving Data Projects' comprehensive and fully detailed guidance with questions and examples makes it a must-have for any data team.

Marilu Lopez, *CEO, SEGDA and author of 'Data Strategies for Data Governance'*

Many business creatives are challenged by numbers and using data to drive their projects. Christine has hacked the code for those who need to understand and utilise data in their work, but don't naturally understand the connection and power of numbers. If you are just starting out in your career or are new to the data landscape, this book will make the connection for you, setting you apart as a valuable team player and allow you to hit the ball out of the park on project after project!

Debra McCarver, *Section Instructor, Carson College of Business*

I am always unexpectedly surprised to learn so much in so short a time. This is packed with solid information that should be in all data professional's toolkits. This is especially true for PMP Certificate holders as the impacts are profound. Insights gained from this will help even seasoned professionals to better understand what they are up against.

Peter Aiken, *Associate Professor of Information Systems,*
Virginia Commonwealth University

DRIVING DATA PROJECTS

BCS, THE CHARTERED INSTITUTE FOR IT

BCS, The Chartered Institute for IT, is committed to making IT good for society. We use the power of our network to bring about positive, tangible change. We champion the global IT profession and the interests of individuals engaged in that profession, for the benefit of all.

Exchanging IT expertise and knowledge
The Institute fosters links between experts from industry, academia and business to promote new thinking, education and knowledge sharing.

Supporting practitioners
Through continuing professional development and a series of respected IT qualifications, the Institute seeks to promote professional practice tuned to the demands of business. It provides practical support and information services to its members and volunteer communities around the world.

Setting standards and frameworks
The Institute collaborates with government, industry and relevant bodies to establish good working practices, codes of conduct, skills frameworks and common standards. It also offers a range of consultancy services to employers to help them adopt best practice.

Become a member
More than 70,000 people, including students, teachers, professionals and practitioners, enjoy the benefits of BCS membership. These include access to an international community, invitations to a roster of local and national events, career development tools and a quarterly thought-leadership magazine. Visit www.bcs.org to find out more.

Further information
BCS, The Chartered Institute for IT,
3 Newbridge Square,
Swindon, SN1 1BY, United Kingdom.
T +44 (0) 1793 417 417
(Monday to Friday, 09:00 to 17:00 UK time)
www.bcs.org/contact

http://shop.bcs.org/
publishing@bcs.uk

DRIVING DATA PROJECTS
A comprehensive guide

By Christine Haskell

Published by BCS Learning and Development Ltd, a wholly owned subsidiary of BCS, The Chartered Institute for IT, 3 Newbridge Square, Swindon, SN1 1BY, UK.
www.bcs.org

Paperback ISBN: 978-1-78017-6239
PDF ISBN: 978-1-78017-6246
ePUB ISBN: 978-1-78017-6253

Ebook available

British Cataloguing in Publication Data.
A CIP catalogue record for this book is available at the British Library.

Disclaimer:
The views expressed in this book are of the authors and do not necessarily reflect the views of the Institute or BCS Learning and Development Ltd except where explicitly stated as such. Although every care has been taken by the authors and BCS Learning and Development Ltd in the preparation of the publication, no warranty is given by the authors or BCS Learning and Development Ltd as publisher as to the accuracy or completeness of the information contained within it and neither the authors nor BCS Learning and Development Ltd shall be responsible or liable for any loss or damage whatsoever arising by virtue of such information or any instructions or advice contained within this publication or by any of the aforementioned.

All URLs were correct at the time of publication.

Publisher's acknowledgements
Reviewers: Maria Papastathi and Nigel Turner
Publisher: Ian Borthwick
Commissioning editor: Heather Wood
Production manager: Florence Leroy
Project manager: Sunrise Setting Ltd
Copy-editor: Kristy Barker
Proofreader: Annette Parkinson
Critical reviewer: Barbara Eastman
Indexer: David Gaskell
Cover design: Alex Wright
Cover image: Fabian Gysel/iStock
Sales director: Charles Rumball
Typeset by Lapiz Digital Services, Chennai, India

CONTENTS

LIST OF FIGURES, TABLES AND EXHIBITS

ABBREVIATIONS

AI	Artificial Intelligence
BDS	business data stewards
BI	Business Intelligence
BSC	balanced scorecard
CDO	chief data officer
CFO	chief financial officer
CIO	chief information officer
CoPs	communities of practice
CPE	customer partner experience
CRM	customer relationship management
DSS	Decision Support Systems
DW	data warehousing
DWSs	data warehousing specialists
EIS	executive information systems
GDPR	General Data Protection Regulation
HR	human resources
IT	information technology
KPIs	key performance indicators
MDM	master data management
ML	machine learning
PAM	privileged access management
PII	personally identifiable information
PoLP	principle of least privilege
PTO	paid time off
RACI	responsible, accountable, consulted and informed
ROI	return on investment
SDCL	software development lifecycle
SLAs	service level agreements
UX	user experience

ACKNOWLEDGEMENTS

Thanks to the students, clients, mentors and colleagues who made use of these materials as part of my lectures, services or solutions. I am grateful for their feedback, encouragement and contributions.

At the risk of missing someone, I will call out a few people I'd like to recognise.

Every book has guardian angels. People who come to the author's aid when needed and help light the way. Danette McGilvray, Tom Redman, Tony Shaw, Chad Richeson and Andy Cotgreave: thank you for being a sounding board for ideas, allowing me to talk through an idea, providing honest feedback, encouragement and contributions, or for being available for those 'quick' questions that are never quick.

Peter Block, Virginia Eubanks, Mando Rotman, Tom Jongen, Fabrizio Lecci, Mattia Ciollaro, William Koenders, Liana Rivas, Jennifer Tucker, Jason Korman and Shubham Bhardwaj for generously sharing their expertise and material, making this book stronger as a result.

Shelley Roberts, Tracey Tomassi, Rebecka June, Amy Gillespie, Degan Walters, Paula Land, Jen Olson, Melissa Garcia Ortiz, Kristin Flandreau, Bobbi Young, Alli Besl, Tina Qunell, Zena Filice, Lianna Appelt, Michael Hetrick, Lovekesh Babbar, Katherine von Jan and Bethany Niese whose encouragement during the peaks and valleys of this journey helped motivate me.

Kaitlyn Halamuda, Dileepa Prabhakar, Christine Gibbons, Joseph Taylor and the anonymous BCS reviewers, for their hours of helpful reviewing and constructive feedback. Your informed, constructive eyes helped keep me on track.

Thanks to those who lead and work in the professional associations, providing me a forum to teach or publish, or offer under-the-hood advice or support. I cannot name everyone here, but I would like to call out Tony Shaw (DataVersity) and Peter Aiken (DAMA) for their support.

To the students and co-faculty of the Carson School of Business at Washington State University, who inspired this manuscript. To those who have sat through my courses and put some of these tools into practice or supported ideas as sponsors, project managers, change managers or data practitioners across a variety of sectors – thank you for sharing your wins, opportunities, challenges and barriers for the benefit of everyone's learning.

Thank you to Charles Rumball, Ian Borthwick and Heather Wood at BCS, who read the initial proposal and detailed manuscript. To Florence Leroy, Sharon Nickels and the countless others I don't know about who contributed to the editing, graphics and typesetting.

And to Steve Banfield, my partner, best friend and the best advocate anyone could hope for. His unwavering support and encouragement, even when I began to doubt, have been the most profound gift one can receive.

FOREWORD

Enterprise data projects are incredibly complex. They typically involve many teams, technologies and business processes, impacting an array of stakeholders such as internal users, partners, suppliers and end customers. They must take into account user requirements as well as privacy and security. They can take considerable time to complete, yet often start only after their need is overdue. It's no coincidence that many data leaders refer to their projects as 'trying to change the wings on a plane in midair'. I agree with this sentiment and would add that sometimes the plane is on fire.

When I worked with Christine at a prominent technology company in the early 2000s, we didn't know we were working on a problem that would reshape the future. We were part of a small team trying to coordinate and streamline metrics and scorecards to enable fast, high-quality decisions. We were intending to be merely data consumers but instead got sucked into the guts of the machine. What we found fascinated us – a discipline that combined art *and* science, had a measurable business impact and saw new challenges emerge every day. The profession was outgrowing its prior approaches and needed new solutions. It was an exciting time.

Fast forward to the present, and a lot has changed. Machine learning and data science are now commonplace at companies. Public clouds allow data professionals instant access to the latest technologies with fewer limitations on storage. Generative AI has burst onto the scene with the potential to reshape companies and industries, with data as its lifeblood.

While data is more important than ever, what hasn't changed is the difficulty of driving enterprise data projects. In fact, the act of designing, implementing and operating a data programme and platform is perhaps more difficult than ever. Despite data volume and variety continuing to expand, budgets are often flat or even down. Data projects are still not treated as top priority at the C-level. Data-forward cultures have been slow to develop, and enterprises still use only a fraction of the data they have available. Especially compared to software projects, data projects don't often get the level of rigour they need. I worked alongside Christine in several roles as we learned these lessons, some more painful than others. We hadn't yet learned how to distinguish signal from noise, when it came to meeting user requirements, making trade-offs in capabilities or evaluating technologies. In this book, the signals are isolated and explained. Christine's case studies, examples and metaphors become guideposts, inviting readers to observe, decide and take action in their own organisations; to recognise the risk signals and mitigate them; and sometimes to get out of their own way.

The knowledge in this book reflects not only many years of real data work done by real people at successful companies but also Christine's distillation of the factors that make data projects successful, based on her many years of enterprise experience. I wish I had had this book 20 years ago, but you have the opportunity to use these lessons as your starting point. And these opportunities will be vast, as emerging technologies like AI will make data even more valuable and your role as a data leader even more important.

It's a great time to be a data professional. The world is coming to our doorstep. But we all have to step up our game.

Chad Richeson
Founder, Firebrand AI

INTRODUCTION

The terminology and methodologies for describing and managing data processes change every 10–20 years. Contemporary data strategies and job profiles now integrate terms like 'data analytics', 'artificial intelligence' and 'machine learning', replacing the previous norm of 'business intelligence' and 'data science'. The emergence of information science degrees globally is relatively recent, aligning with the swift changes in the world and technological progress.

With the exponential changes happening in the world and the pace of technological advances, the classic trinity through which an organisation is improved – the Three Ps: people, process and platforms (technology) – is changing. This premise is so ingrained in business and in graduate programmes that each has its own methodology: people (change management methods such as ProSci), process (LEAN, 6Sigma and so on) and platforms (with a variety of technical platform management techniques). However, 'data' has become a legitimate, *fourth* distinct discipline worthy of consideration. To date, it is categorised into data science, data engineering, and the like. This has merit; still, there is as yet no current standard methodology to help uplift the general population's data skills. But people are starting to talk about it.

Many employees seek out or are thrust into a series of responsibilities in data management for which there is little formal training. How they engage with data in those roles impacts the privacy and security of consumer data and the overall risk to the company's bottom line. The problem? They aren't quite sure how data works or how to drive data projects – not really.

Today, almost all projects involve data to some degree, yet the data aspect is not adequately addressed. *All technology projects are data projects*. There is a general lack of understanding of the data supply chain (and our responsibilities and accountabilities in that process), and we must improve our project and change management rigour. In companies where management is not prepared, trained or incentivised to nurture data cultures, there are some basic frameworks and methodologies to rely upon. In companies where data-driven processes are supported and encouraged, teams can learn to improvise, create tools, adapt to change and prove value. Either way, organisations are never starting from nothing; everyone is trying to find a path forward.

While there are endless certifications in data tools, not everyone needs them to understand data enough to drive an initiative. Until data becomes a legitimate business discipline, certifications in various project or product methodologies can help. But paying to learn material central to passing a certification exam rather than how to apply skills directly can feel more regressive than progressive. Certifications are, however, a means towards exposure to different philosophies on how to drive work forward. They provide environments of highly focused learning and support not always available in the workplace. The most valuable teachers of how to drive work forward are those people and cultures that resist our efforts in driving data projects. If we observe them carefully and learn from them, they teach us how to adapt our methods and tools to our environment, not the textbook domain (where everything works out). If approached with humility, certifications are not the end goal. Instead, they can be a way of gaining exposure to concepts, tools and templates to help us develop a roadmap for a constantly evolving learning journey and to collaborate more effectively with other disciplines.

The answers to my own questions about driving data projects forward were found in my innate curiosity, many books, several certifications and a broad spectrum of experiences (others' and my own). I learned that I could not fail if I knew how to adapt my toolset and become more dynamic with my skills. This insight was very liberating. I started learning to adapt tools from certification methodologies to the needs and constraints of my unique environments. I focused on what worked (and why), learning about data concepts I genuinely cared about (such as the data supply chain), and gained experiences that took me far from my comfort zone (such as solving business problems through data management, writing this book and teaching graduate school). Theory helps to teach us how to think well but is often not grounded in the real world. It's the wisdom of applied practice that everyone covets the most.

The suggestions in this book are presented only as *suggestions* for understanding the data supply chain (and your responsibilities within it) and an approach to managing data projects more constructively with business stakeholders. The tools, templates and ideas in this book are here to inspire and ground you when you are looking for the next step on your path. I invite you to find something that sparks your interest and form your own experiments on how to best drive change in your organisation. These suggestions come from practical knowledge and experience gained from years of trial and error in organisational cultures across different sectors. Sometimes, it can be satisfying to learn battle-tested ideas from someone who spent their career trying new ideas every day and is willing to share their results.

Over the past 20 years, observing and leading transformational initiatives in organisations has enabled me to build an approach to driving work forward – a way of observing, collaborating and adapting – using tools and ideas from multiple methodologies. The invitation extended here is for you to learn and adapt these ideas to your environment, making them more relevant to your circumstances.

The ideas in this book are not all my own, nor do I deserve credit for them. They come from long-standing guilds such as the Project Management Institute, ProSci Change Management, BCS, The Chartered Institute for IT, the Data Management Association and many others before me. The opportunity here is that we should develop the skills necessary to examine our approaches to using data and the many ethical, analytical and technical problems that impact our lives, organisations and society. Hopefully, the ideas and examples in this book provide a perspective on and example of what finding a path forward – knowing what the next step is – can enable.

The timeless, broad ideas and practical applications presented in this book are offered for my students, teams and colleagues looking to develop a more rigorous approach forward. In developing this work, I hope to inspire others to learn about their role *and impact* in the data supply chain with enthusiasm, clarity and confidence, helping to make their skill development continuous and immediately impactful.

It is ironic that, having gained English literature and industrial psychology degrees, it was data and information technology that gave me a voice and a great career. My hope is that you bring your gifts and perspective to your role in data to help further unlock its potential – and yours!

BE INTENTIONAL. MAKE AN IMPACT.
Christine Haskell, PhD
Seattle, WA

A WORD ABOUT THE BOOK COVER

Generally speaking, I wouldn't comment much on a book's cover and am the last person to give creative direction. I spend most of my time synthesising information and simplifying complexity to enable decisions, and I envy the rich vocabulary of an effective marketer.

The cover shows the interior cockpit of an airplane, with a pilot and copilot 'driving'. A high-tech cockpit provides tools to keep us safe. It conveys the strategic (and necessary) role of data to drive the success of an operation, and not just data for data's sake. It shows how to present data in a way that can be most easily consumed by the user and utilised to make decisions. Flying planes is not without risk, but as so many lives depend on the safety of the flight, the pilot oversees the flight to reduce risk as much as possible. As such, I was okay to move forward with it.

What was missing, though, was an image that could convey the complex knowledge needed to fly the plane. Pilots need to have both flying *and* mechanic knowledge. They need to know what is happening with the plane as much as they need to understand the dynamics of the plane and the environment (that is, the flight itself). Understanding how the wing angles operate during certain weather events lets the pilot use the aircraft's features precisely, saving time and fuel while managing safety. Pilots are supplying a service to passengers, who are given basic rules of engaging with the plane during the flight.

Many data professionals started out thinking they would be consumers of data services. However, they quickly realised that consuming data meant understanding how it's defined and coded, where it comes from and how it behaves over time – the mechanics. The forces requiring us to understand the underlying mechanics of our business strategy and performance become 'business intelligence', now called 'data and analytics', and inform us how to manage our business strategies more precisely.

Data management hasn't changed in the past hundred years or even in the past fifty. It's still about getting from point A to point B efficiently, cost-effectively and safely. Mastering the basic mechanics of data management is critical for a data team's success, and everyone else's.

What *has* changed is that technologies such as Artificial Intelligence (AI) have emphasised data management as a discipline. The speed of decision-making and the consequences of (poor) data management have increased. With AI and machine learning, data teams and organisations are now powered by high-tech cockpits, requiring everyone, at every level, to raise their skill level. Where before, people might have been able to consume reports unquestioningly – much like airline passengers – today they must have a rudimentary understanding of how data works to use it effectively and responsibly.

If the organisation or the teams tasked with managing data don't understand how data works, lead data as one-and-done projects or don't align data projects as initiatives that support a broader business strategy, AI will only magnify poorly asked questions, existing quality problems and poor project execution.

The ongoing pressure to gain a deeper understanding of data will persist, especially when unchecked AI outcomes are applied and users rely on the assumption of high data quality without thoroughly verifying lineage or confirming the accuracy of results. This dynamic enables that very (incorrect) data to continue to train the AI algorithm. Now more than ever, we need all professionals using data who can tell the difference. Consumers of AI data services must continue to shine a very harsh light on the quality of data services.

With powerful technologies such as AI, data management professionals at every level must have basic driving *and* mechanic skills – and the ability to switch dynamically between the two roles.

INTENDED AUDIENCE

This book will help anyone responsible for or who cares about deeper data team and business strategy integration in their organisation. However, individuals in different roles will find different aspects of the book useful.

Early career and those new to data

This applies to members of a team or staff who do the hands-on work of managing projects that have platform or data components as part of their daily responsibilities. You might be:

- just finishing school and looking to apply what you have learned in the 'real world';
- changing careers and seeking the skills and experience needed to achieve long-term success in the field;
- working in a business function where data has been considered an afterthought or where the function feels like it is trying to 'catch up' in terms of understanding how to work with data.

Data should not be an afterthought in any modern business function. However, in some cases, due to historical practices or a lack of awareness about the importance of data, it might be treated as a less prominent consideration in certain business functions. Here are a few areas where data has historically been treated as an afterthought:

- *Human resources (HR)*: In the past, HR departments may have focused more on administrative tasks and compliance, with less emphasis on data-driven decision-making. However, modern HR practices increasingly rely on data for talent acquisition, performance management and workforce planning.

- *Customer service*: While customer service has always been a data-intensive function, some organisations may have neglected to leverage data effectively to improve customer experiences. Nowadays, businesses are more inclined to use data analytics to understand customer needs and preferences.

- *Supply chain management*: Historically, supply chain management may have been more focused on logistics and cost efficiency than on data analysis. However, data is becoming increasingly important in optimising supply chains for efficiency, visibility and resilience.

- *Facilities management*: Facilities management often involves maintaining physical infrastructure and assets. Data, such as maintenance records and energy consumption data, can be underutilised in optimising facility operations. Modern practices include using data to enhance sustainability and reduce operational costs.

- *Marketing (in some cases)*: While marketing has always used data for targeting and measuring campaign success, there has been a shift towards data-driven marketing strategies. Some organisations may still rely on traditional marketing methods without fully harnessing the power of data analytics.

These areas are evolving quickly, and many businesses are recognising the importance of data-driven decision-making across *all* functions. As a result, organisations increasingly invest in data analytics, artificial intelligence and data science to make data a central part of their decision-making processes in every business function, *requiring all functions to understand how data works*.

Data or technical practitioners

Data professionals frequently engage with various business functions when stakeholders require data or reports. Meeting data requests while lacking context within a business strategy is a formidable challenge in terms of accuracy and consistency. Consequently, data professionals can increase their effectiveness by gaining a better understanding of business strategy, enabling them to align their data-related efforts with their organisation's overarching goals and objectives. As much as data and technology teams might want to help business stakeholders, or as much as the business might want to leverage technology teams, trust is not a given. Mutual distrust between business stakeholders and information technology (IT) teams can arise due to a combination of factors, often stemming from miscommunication, differing priorities and past experiences. Here are some reasons for this mutual distrust:

- *Communication barriers*: Technical jargon and complex terminology IT professionals use can be difficult for business stakeholders to understand. This communication gap can create a sense of exclusion and frustration.

- *Misaligned priorities*: Business stakeholders typically focus on achieving business objectives, while IT teams may prioritise technical considerations. Misalignment of priorities can lead to conflicts and mistrust.

- *Lack of transparency*: When IT teams do not provide clear explanations or updates on project progress, business stakeholders may perceive a lack of transparency and assume the worst.

- *Budget and resource issues*: Business stakeholders may question the allocation of resources and budget for IT projects if they do not see a clear return on investment or if projects frequently exceed their budgets.

- *Missed deadlines*: Frequent delays in project delivery can erode trust. When IT teams consistently fail to meet deadlines, business stakeholders may doubt their ability to execute effectively.

- *Scope creep*: Expanding project scope without proper communication can lead to misunderstandings and scepticism. Business stakeholders may think that IT teams are not managing projects efficiently.

- *Past negative experiences*: If business stakeholders have had previous negative experiences with IT projects, such as failures, budget overruns, or poor communication, they are likely to approach future collaborations with scepticism.

- *Perceived inflexibility*: If IT teams resist changes proposed by business stakeholders based on technical feasibility or security concerns, resistance can result in conflict and mistrust.

- *Security concerns*: Data breaches or security incidents can lead to substantial mistrust if IT teams are perceived as not adequately protecting sensitive information.

- *Failure to understand business needs*: IT teams must fully grasp the business's core objectives to provide effective solutions. When business needs are not understood or addressed, stakeholders may become sceptical.

- *Inadequate problem-solving*: When IT teams do not efficiently address technical issues or provide solutions that align with business objectives, stakeholders may lose confidence in their ability to solve problems.

- *Cultural differences*: Differences in culture, work styles or perspectives between IT and business teams can create distrust due to a lack of alignment and mutual understanding.

Building the most constructive bridge between data teams and business stakeholders requires a solid understanding of the digital supply chain (and our accountabilities within it), and a rigorous approach to project and change management. IT teams and business stakeholders need to prioritise open and transparent communication. They should work collaboratively to understand each other's needs, expectations and constraints. Building trust requires both parties to commit to shared goals and responsibilities, acknowledge past issues and work together to find solutions that benefit the organisation as a whole.

Everyone

Everyone must gain a basic understanding of the how data travels though an organisation – and of their *individual* accountability in that process. They must also understand how data projects connect to one another, directly enabling business strategy. Such awareness can positively impact individuals, organisations and society as a whole.

Here are some key advantages to these two very basic data awareness concepts:

- *Informed decision-making*: Data awareness empowers individuals to make informed decisions in various aspects of life, from personal finances to healthcare choices. In a business context, it enables better decision-making at all levels, from executives to front-line employees.

- *Improved problem-solving*: Data awareness equips individuals with the skills to analyse and solve complex problems. This is invaluable in both professional and personal settings.

- *Increased efficiency*: Data-aware individuals can work more efficiently. In a workplace this translates to optimised processes, reduced inefficiencies and improved productivity.

- *Better communication*: Data awareness fosters the ability to convey ideas and information effectively using data. This is essential for persuading others, presenting findings and collaborating with colleagues.

- *Competitive advantage*: Organisations with a data-aware workforce are more competitive. They can adapt to changing market conditions, identify opportunities and respond to challenges more effectively.

- *Innovation*: Data awareness fuels innovation. When individuals understand how to work with data, they can identify new ideas and solutions, leading to innovation and creativity.

- *Data-driven culture*: Promoting data awareness within an organisation encourages a data-driven culture. This means that decisions are based on evidence and data, rather than gut feelings or intuition.

- *Accountability and transparency*: Data awareness fosters accountability and transparency. When individuals and organisations are able to back their claims and decisions with data, it builds trust with stakeholders.

- *Personal growth*: Developing data awareness can enhance personal and professional growth. It can open up new career opportunities and improve job security in an increasingly data-driven world.

- *Data ethics*: Data awareness also includes understanding the ethical considerations related to data, privacy and security. This helps individuals and organisations navigate the ethical challenges of collecting and using data responsibly.

- *Resource allocation*: Both individuals and organisations can make better decisions about resource allocation when they understand data. This includes allocating time, budget and personnel more effectively.

- *Global problem-solving*: Many of the world's most pressing challenges, such as climate change and public health crises, require data-driven solutions. A more data-aware global population can contribute to solving these challenges.

- *Policy and governance*: Data-aware citizens are better equipped to engage in discussions about policy and governance, leading to more informed and effective policies.

- *Education and research*: Data awareness is crucial for students and researchers across various disciplines. It enables them to conduct research, analyse findings and make evidence-based conclusions.

- *Data-driven innovation*: Data literacy is a catalyst for data-driven innovation. It enables individuals and organisations to harness the power of data analytics, artificial intelligence and machine learning for positive change.

In summary, understanding the basics for how data becomes an asset is a fundamental skill in the 21st century that can benefit individuals, organisations, and society in numerous ways. It enhances decision-making, problem-solving, communication and overall efficiency, leading to better outcomes in both personal and professional spheres.

HOW TO USE THIS RESOURCE

This guide is a data project primer – an overview of what data is, the supply chain, its use today, what distinguishes it from traditional operations. Read this introduction to understand the context in which your data projects will fit and the guidelines critical to your success.

Familiarise yourself with the data supply journey in Chapter 1. Understand the human and machine perspectives, gain awareness of where you are in that journey and consider what accountabilities you have towards ensuring data quality. Become familiar with the blended methodology in Chapters 2–6 by initially skimming the table of contents and by skimming the book from beginning to end. This reflection will orient you to what is available and where it fits in the book.

Driving Data Projects explores the basics of the data supply chain and a four-phase process for successfully implementing and refining data projects (see Figure 0.1). This guide equips you with the knowledge and skills to identify project requirements, gain and manage sponsorship, oversee the project and continuously improve strategies. Following key principles, you'll effectively address specific business needs, align resources efficiently and foster ongoing organisational change through mutual learning experiences.

Figure 0.1 Four-phase process for driving and fine-tuning data projects

Scope the project	Determine resources	Manage the work	Tune the change
Focus on the right projects. Identify the things that are important to your project and are a good match for your organisation's data management efforts.	Confirm the right resources. Locate and partner with data providers and consumers (internal and external).	Ruthlessly prioritise. Don't boil the ocean. Select the things that will make meaningful and documented impact.	Leverage your work by sustaining what you have built so that data as a service become an impactful tool in your toolkit and is fully integrated across your organisation.

Each chapter describes the first steps and advanced tactics to drive data projects. They are presented separately, so you know what to try as a beginner and what to revisit when you have more experience. It is important to embrace what's feasible, never losing sight of the big gains possible when prioritising data as a service. By adopting the practices in this guide as a potential path through the chaos, you could generate significant cost savings each year while increasing the value of data as an asset for your organisation. You could also improve how you manage internal and external resources, stakeholders and partners and your ability to handle various data projects. Using ongoing data requests from the business as an opportunity to intentionally integrate data as a core component of business strategy, you can upskill teams, increase organisation-wide influence and become a far more effective, consultative partner to your business peers.

Phase One defines a step-by-step approach to scoping your data project. You will select a few options for data projects based on your multiyear goals and vet these opportunities against four criteria for successful projects. The book will walk through and share examples of simple data projects. It is important to create a *scope statement* that clearly communicates what you want and need from the people working on your team. Details are provided on how to write a concrete scope statement.

Phase Two focuses on determining resources. A blended view of project management and change management methods is explored to help you determine how to craft the best path forward. As the data project lead, your relationship with the executive sponsor and the stakeholder/working group is explored. Various project team structures are presented, with assessment tools to help the project manager determine enablement needs and take relevant steps to resolve knowledge gaps. A data project stakeholder map is provided to help you think through how to leverage relationships across your organisation. Project and change tools help the project lead build supporting teams and effectively align with and leverage the group members.

Phase Three focuses on successfully managing the work, starting with five key ideas for fine-tuning data initiatives. Get everyone on the same page, ensure the project stays on track and ensure the human–machine contributions are clear. The focus is on a team-based project for a single initiative because it is one of the most complex types of cross-functional sponsorships, requiring multiple people from different functions (and sometimes external partners in other sectors) and backgrounds working towards a single goal. Confirming sponsorship for this type of project guarantees you can manage any type of data management project. Prepare for the project over five phases: prepare and plan; kick-off; discover; build, deliver and implement the data transformation; evaluate and celebrate success. Tools and strategies are introduced to support every stage.

Phase Four suggests ways to fine-tune your efforts to manage data transformation in a way that strategically increases your capacity. The goal is to build a robust data culture as part of your project delivery, turning your data-as-a-service work into systematic support beyond your foundational data projects. Consider how to weave data as a service into your strategic planning process and staffing plans, ideas for building long-term relationships with stakeholders and other steps to create the right foundation to support your data transformation journey. Review the five principles and four general 'levels' of maturity which *Driving Data Project* organisations use to progress their work, and what you need to do to pass from each level to the next, from observer to skilled.

This book is a *hands-on guide* exposing suggestions for implementing a data project as a step towards a more data-informed culture. It contains questions, worksheets, checklists, case studies and tools designed to help your team brainstorm, build, problem-solve, plan and make decisions about each data project *with less chaos and more intention*. Every tool in this book should be processed working as a team, with input from all levels and roles on the data project. Data projects by themselves don't attract budget and executive attention and support. However, momentum, with enough projects, successful initiatives and traction in solving challenging business problems, does. The guide can also serve as a training resource and introductory overview for those early in their career or those new to data transformation activities and trying to find an initial path. Business stakeholders can also become better data partners once they understand everyone's role in the data supply chain.

1 DATA FOUNDATIONS

THE BASICS

Data has traditionally been managed by a combination of information technology (IT), operations and finance. Over the past 10–15 years, the Chief Data Officer (CDO) role has appeared on the executive scene. While it is not yet a universally used title, the CDO role began with reporting through these functions and is beginning to be considered separately.

The CDO's primary mission, though not always stated directly, is to cultivate data as a strategic asset for the organisation. This premise shifts the emphasis on merely embracing the most efficient approach to managing ad hoc requests for a data pipeline to a discrete group of stakeholders, to prioritising data (and perhaps even the internal mechanisms of the data supply chain) as a separate asset for the *whole* organisation. It also speaks to data's integrated and entangled nature – every function depends on it. To better prioritise data across the organisation, the CDO role needs *every* employee to work with data effectively across all functions. This need means that working with data is no longer just a specialist skill; it's a critical workforce skill. The data team must take the lead, using every project to educate stakeholders about the needs, implications and consequences of managing data. They must take these projects as an opportunity to explain how data works and help stakeholders take a more active role in driving accountability and reducing bias.

This guide attempts to help data teams become more effective drivers of data projects and initiatives by illustrating the data supply chain from the perspectives of the business *and* data teams. It also offers a blended methodology for driving and aligning data work with business strategy. The guide focuses on the mechanics of nurturing data projects through to completion. Nurturing data is about creating a trusted data foundation, stewarding the analytics and data science analysis, cultivating capability, driving innovation and working closely with business stakeholders to monetise the data in the most ethical fashion. Since many technical teams find connecting to business stakeholders and processes difficult, this guide will help them level up into the business realm by leveraging integrated project and change management approaches to drive transformation. The IT and operations functions have been (somewhat clumsily) referred to as the organisation's 'plumbers' because data has been considered a standard utility. This narrower value lens emphasises efficiency, productivity and cost reduction. While those values matter, the role of the CDO has elevated the importance of data in the organisation by focusing on the overall *business value* of data services.

Today, accounting for what data is needed and how it will be used to drive decisions must become an essential aspect of *every* functional strategy in the organisation.

Because every function needs data, the CDO must partner intentionally and strategically with its peers to successfully integrate data across the company while determining the architecture required to support those strategies. Viewed from this perspective, the metaphor for managing data changes from that of mere utility to more of a natural resource, which we will discuss later.

This guide evolved in response to a simple question from my graduate students: *How do I manage a data project from start to end, regardless of organisation, process or data management maturity levels?* Relevant situations might be:

- A project manager in the travel sector was just told to 'figure out executive measures for the division'.
- An operations lead in finance needs to determine an executive scorecard monitoring a migration from an internal platform to SAP.
- A programme manager is tasked with implementing a customer relationship management (CRM) tool.
- An industrial psychologist in the technology sector must gather data for rapid insights for briefings with executive management. These insights will inform decisions on budget, programming and strategy for the division.
- A leader in the retail sector must uplevel her team's project management skills to partner more efficiently with data teams.
- A product lead in the aerospace sector must develop a report to monitor intrapreneurial efforts, providing more efficient maintenance services.
- A data engineer must track storage spend with a third-party vendor.

Many people don't think of these as data projects. They think of them as technology projects. Leaders invest in a technology and then try to retrofit everything around that decision. Many technology decisions impact or, in some cases, determine data strategies for an organisation but aren't recognised as such.

I, too, have faced requests such as these in every organisation I've served, and hear from other data management professionals presented with the same issues, challenges and concerns. Requests for data often come to us with little context, immediate deadlines and expectations for fast fixes. The work might not always appear to align with the current goals of the data team. When taking on data projects, most students and data professionals feel overwhelmed. Where to start? How do you think through the problem from all angles?

Eventually, it becomes a creative puzzle to assemble. Data projects help highlight dependencies between teams that can inform decisions and interdependent strategies. There is often a need for reliable data sources, identifying someone who can help aggregate and prepare the data and another resource who can confirm accurate insights. From there, data storage, sharing and monetising become concerns. Over time, a larger picture of interconnected activities and systems – the data stack – becomes visible.

Data maturity is a model commonly used to describe how companies, groups or individuals advanced through stages of data analysis over time. It helps data professionals diagnose where they are and where they want to be. Low maturity is typically at the bottom left (see Figure 1.1) and focuses on manual efforts, establishing initial processes and understanding what measures will impact the business. High maturity is at the top right (see Figure 1.1), focused on predictive data, artificial intelligence and machine learning. A model such as this can help us identify where our organisation is. Many self-identify their current efforts as being left of centre (see Figure 1.1) and lacking maturity. It is here that they often judge themselves and their organisations harshly. They shouldn't.

Figure 1.1 Overview of the Maturity Model for Data and Analytics (Source: Gartner, Gartner Survey Shows Organizations Are Slow to Advance in Data and Analytics, 5 February 2018 (https://www.gartner.com/en/newsroom/press-releases/2018-02-05-gartner-survey-shows-organizations-are-slow-to-advance-in-data-and-analytics))

Level 1 Basic	Level 2 Opportunistic	Level 3 Systematic	Level 4 Differentiating	Level 5 Transformational
• Data is not exploited, it is used • D&A is managed in silos • People argue about whose data is correct	• IT attempts to formalize information availability requirements • Progress is hampered by culture; inconsistent incentives	• Different content types are still treated differently • Strategy and vision formed (five pages)	• Executives champion and communicate best practices	• D&A is central to business strategy
	• Organizational barriers and lack of leadership	• Agile emerges • Exogenous data sources are readily integrated	• Business-led/driven, with CDO • D&A is an indispensable fuel for performance and innovation, and linked across programs	• Data value influences investments • Strategy and execution aligned and continually improved
• Analysis is ad hoc • Spreadsheet and information firefighting • Transactional	• Strategy is over 100 pages; not business-relevant • Data quality and insight efforts, but still in silos	• Business executives become D&A champions	• Program mgmt.. mentality for ongoing synergy • Link to outcome and data used for ROI	• Outside-in perspective • CDO sits on board

D&A = data and analytics; ROI = return on investment

© 2017 Gartner, Inc.

Data transformation and related data projects like the ones listed above are not new. Yet, for every organisation, they are a challenge. Many organisations – even large, sophisticated companies – engage with data on a highly ad hoc basis, with little management rigour. This presents two common problems. First, stakeholders find it hard to know whether (and when) the data team will meet their requirements. This ambiguity (and sometimes the turnaround time) encourages them to build their own solutions, which work for a while, until they don't. Independent efforts that run too long without cross-organisational alignment often result in disaster. Second, the data team burns out from overwhelm due to endless requests (often lacking sufficient context) and has no method for prioritising the many conflicting demands on its time (because everything is priority 0). The heroics involved in seeking to meet every ad hoc request lead to burnout, and increased attrition makes the problem worse.

> No matter how small, every data project is an opportunity to seed the movement towards a data-forward organisation – one that culturally treats data as a strategic asset and builds capabilities to put that asset to use, as an aid, for all manner of decisions. Every project can become an example of executing a data management strategy more intentionally with business strategy.

The data needs and challenges presenting themselves today are symptoms of the data maturity level of your organisation. Those challenges feel unique and special because we live in our particulars. This book will provide principles and tools to increase your ability to drive meaningful insights, influence business decision-making and amplify your existing effectiveness, from running your first data project to fine-tuning data as an ongoing service.

Digital intelligence

Digital intelligence is a comprehensive set of technical, cognitive, meta-cognitive and socio-emotional competencies grounded in universal moral values and enabling individuals to face the challenges and harness the opportunities of digital life.[1] Additionally, much has been made of data literacy, data fluency, the need for data skills, and the like. These and other data initiatives are examples of an urgent need to empower individuals in their work and lives with new competencies and greater rigour that can help increase readiness for the rapid advance of AI and other technologies in the near future.

Companies are starting to invest millions of dollars in 'data literacy', 'digital skills' and 'digital readiness' programmes across all sectors around the world. However, with limited coordination, such efforts are not evenly distributed or consistently applied, as there is a lack of globally shared principles or a central framework for digital competencies and standards.

Leading data projects has never been easy. Given the frequency and challenge of security threats, privacy concerns and the business's growing demands for timely, accurate data, just trying to manage the day-to-day presents unprecedented complexity for executives. Governments are in crisis. Demographics are shifting. Globalisation and market demands require every individual to understand how to 'speak data' across every function – from creative to technical. At the same time, managing data across a flexible, distributed organisation is becoming an increasingly complex business. More than ever before, we continue to blend on- and offline, virtual and physical, with flexible schedules between home and work.

To lead data projects effectively requires everyone to have a certain level of competency in understanding data (what it is, where it comes from, how it can/should be used and so on). An executive who reviews measures in a business review, an analyst who produces reports or even an engineer who monitors product health is not automatically data-fluent just because they look at data. Table 1.1 shows how the language of a definition shapes the questions asked and skills expected of data consumers. Definitions expand and contract based on the demands and tolerance of the culture in which they operate.

---

1   https://www.dqinstitute.org/wp-content/uploads/2023/03/DQGlobalStandardsReport2019.pdf

Table 1.1 How definitions influence competency levels

Source	Definition	Questions emphasised	Suggested Skills
Company (private sector)	Data literacy is the ability to explore, understand and communicate with data.	Can users effectively navigate tools necessary to interpret and create basic visualisations? Can they create effective data stories to drive discussion?	Tool usage, decision-making, storytelling, influencing
Wolfe (academic)	Data literacy is the ability to ask and answer real-world questions from large and small datasets through an inquiry process, with consideration of ethical use of data.[2]	Can we do 'x' with the data, versus should we do 'x' with the data?	Skills from tool usage, decision-making, storytelling, influencing, plus critical thinking, ethical concerns (i.e. accountability, inclusiveness, reliability and safety, fairness, transparency, privacy and security) of data usage
Gartner (advisory and research firm)	Data literacy is the ability to read, write and communicate data in context, including an understanding of data sources and constructs, analytical methods and techniques applied, and the ability to describe the use-case application and resulting value[3]	Do data consumers understand where data came from; can they challenge its structure, quality, and meaning? Can users link measures to specific business problems and outcomes?	General focus in three key areas: a shift in mindset (attitudes and beliefs about data), clarity towards a common language (business terms, data terms, analytical) and skill development (how do we think, engage with others and take action)

If the definition is scoped to target only those who navigate tools for visualisations or expect every employee to create sophisticated visualisations, it leaves many people out. *Everyone* at all levels must understand how to use data effectively and responsibly to influence decisions, take action and drive behaviours – all in service of the organisation's mission. Driving data projects is not only about presenting business requirements (e.g. figuring out executive measures for the division, monitoring an SAP migration, etc.). When considering these requests from the business, we must educate and collaborate

---

2 https://www.researchgate.net/publication/347531630_Creating_an_Understanding_of_Data_Literacy_for_a_Data-driven_Society

3 Gartner IT Glossary, 'Data Literacy', 24 November 2023, https://www.gartner.com/en/information-technology/glossary/data-literacy. GARTNER is a registered trademark of Gartner, Inc. and/or its affiliates and is used herein with permission. All rights reserved.

with the business to first do no harm with the data we collect. We must be mindful and follow regulations to meet broader community expectations. Increased transparency and monitoring of storage costs will help bring awareness to ongoing usage patterns. Under tight constraints to deliver, developers must take calculated shortcuts to complete projects quickly. However, over time, these shortcuts will accumulate into 'debt', at least some of which will need to be 'paid back'. New generations of employees are demanding more from their employers. On top of that, we must provide and manage all this new infrastructure while being told we cannot spend a budget on overheads equivalent to that of our functional peers.

Funding these demands is another matter. Data transformation – the kind that helps inform the organisation's operating model and truly helps business efficiency – cannot happen without considering how to manage data aligned with these goals. Yet, many leaders inhibit their organisation's data maturity by prohibiting access to data, supporting siloed efforts instead of driving cross-team or cross-functional alignment and discouraging appropriate data sharing among data owners. They place demands on data teams to reduce risk and save costs while not acknowledging that the people, processes and platforms to manage data are not a free service. Data teams cannot continually absorb work.

One behaviour that holds more organisations back from increasing their data maturity is leaders who don't continually model the use of data in their decision-making. It seems so basic (because it is). People mimic what they see modelled in their leaders. Leaders at every level can model what they want to see by asking simple questions like, *'When we say "data ethics", does everyone know what we mean?'* or *'Do we all agree on the calculation for this index, and does it incentivise the right behaviours?'*

A lack of maturity shows up when well-intentioned efforts only deliver temporary benefits. Constant ad hoc requests emphasising cost savings undermine efforts to maximise business value from data, analytics and insights – when the demand for insights is at an unprecedented level. It also forces a return to the status quo of constant ad hoc management.

Why data and analytics maturity curves should come with a warning

According to Gartner, there are five levels of data and analytics maturity: basic, opportunistic, systematic, differentiating and transformational. Figure 1.1 illustrates the 2018 Gartner data and analytics maturity curve, which has been used successfully as a diagnostic tool for more than a decade. It is a standard because it lays out actual-to-desired organisational performance levels for data management activities. It provides specific behaviours mapping to those levels. There is a 'you are here' aspect to the visual, enabling easy self-identification of an organisation's current level. Knowing where we are provides a clearer sense of where we want to go.

At the same time, maturity curves can be misleading. All bands are presented as the same – as if each level of maturity required the same amount of time, budget and resources to invest and achieve. *That could not be further from the truth.*

To achieve Level 5, for example, the organisation must be prepared to invest in and deliver an architecture enabling predictive analytics – a barrier for many. There must be a persistent staging layer, a data warehouse, a model of metrics to run the business, different ways to view the business, a single source of truth and an experimentation layer. Figure 1.2 shows what maturity might look like if we looked at it more realistically.[4] Table 1.2 illustrates the relationship between complexity and business value as data analytics matures.

Data projects that are meant to last demand mindset transformation. It is not solely about developing and accessing information buried in text. It is about developing integrated strategies that (1) implement and decide the right processes, business models and domains and (2) cultivate a sustainable cultural transformation that will enable accurate insights and maintain the corporate asset. Data projects built to scale – the focus of this book – align and refine the ongoing tuning process to support informed, ongoing decision-making across the business.

Maturity shows up in the level of questions being asked and the underlying technology invested to support them by executives. One of the biggest problems with the re-emphasis on increasing data skills among employees is a desire to excitedly advocate jumping straight to the predictive side of the maturity curve. This 'leap' is common for most business stakeholders when they see what is available on the market. However, predictive analytics are unattainable for most organisations without significant upfront investment and absorbing massive risk. More organisations are starting to re-group and reassess their efforts in the middle of the curve. Many still have a way to go, even in the early steps.

Data teams may take one step back for every two steps gained, but the path always leads upward. See Table 1.3 for an example of the mindset journey most organisations experience.

But what of the required data skills of data consumers? How should those grow and evolve along with the skills and deliverables of the data team?

From ad hoc efforts to seeding a revolution

'You can't manage what you don't measure.'

The wisdom of this mantra, attributed to both W. Edwards Deming and Peter Drucker, has guided project methodologies and business strategies for decades. It explains why the need for data as an ongoing service is so important. If we seek to know radically more about our businesses and can directly translate that knowledge into improved decision-making and performance, the need for increasingly nuanced, sensitive and protected data will never go away. Insights from data have helped us increase the sophistication with which we manage customer relationships. We have gone from paper-card loyalty programmes driving knowledge of unique purchases made to small segments of customers, to having so much data on every customer that we can predict what they will buy next, how much and how often.

---

4 Adapted from the Five Stages of Analytics Maturity developed by Tom Davenport and Jeanne Harris in their book *Competing on Analytics: The New Science of Winning* (Harvard University Press 2007) and the DELTA Model developed in 2010 by Tom Davenport, Jeanne Harris and Bob Morison in their book *Analytics at Work: Smarter Decisions, Better Results* (Harvard Business Review Press 2010).

Figure 1.2 Data analytics maturity models and the spectrum of related technologies

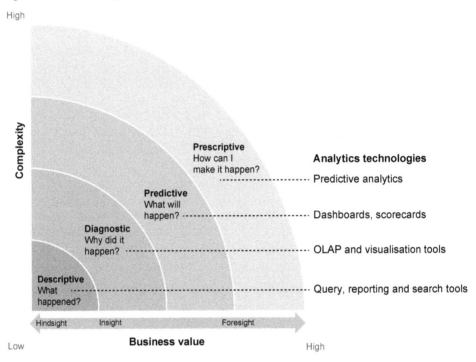

Table 1.2 Data analytics maturity and business value

	Descriptive (Hindsight)	Diagnostic (Insight)	Predictive (Foresight)	Prescriptive (Foresight)
Complexity →	Measures what happened.	Understands what happened.	Predicts what will happen in the future.	Analyses which factor to use to reach an output in a particular scenario.
	Actual financial statements Key performance indicators (KPIs) Headcount reporting	Drills down, filters and uses 80/20 analysis to understand the cause of what happened.	Forecasting (cashflow, sales, etc.) for the next month based on statistical model and adapted after manual review based on the latest business events.	Focuses on finding actionable insights based on the combination of diagnostic and predictive analysis.
	Business value →			

Table 1.3 Sample data mindset journey

Data analytics maturity phases	Poor	Good	Strong	Advanced	Reliable
	Data is inconsistent, unreliable and incomplete.	Difficult for decision makers to access and analyse.	Available and accessible to decision-makers.	Includes data science and statistics expertise.	Recommendations for events and outcomes through data and analytics.
Required mindset	• There is no dominant or required mindset. Data is not top of mind.	• Data is the foundation of analytics. • Successful analysis requires consistent and structured data.	• Data and analytics support decisions throughout the organisation. • Decision makers take a data-driven approach.	• A sophisticated mathematical logic of the situation supports critical decisions. • The data-driven approach is highly trusted.	• Analytics informs and enhances human judgement, contributing to more accurate predictions of outcomes (e.g. scenario planning) and events (e.g. attrition).
Required skills (data team, provisioning)	• Data collected and analysed ad hoc, manually.	• Transactional data-entry systems. • Central data repository and controls. • Basic rules for data collection.	• Technology for disseminating data and analytics in an organisation (e.g. reporting tools, platforms). • Analyst dedicated to internal clients.	• Advanced-analytics specialists (e.g. data scientists, statisticians). • Advanced statistical tools (e.g. R, Python, SPSS) for analysis.	• 'Very big data' – high volume and high reliability. • Deep expertise in predictive analytics (e.g. supervised or unsupervised machine learning).

The sheer volume, velocity and variety of data acquired, transformed and consumed by organisations doesn't guarantee better use of the data; it just guarantees *more* data. The data available are often unstructured. Unorganised data is unwieldy (and costly) data. This behaviour creates many undiscovered insights hidden in large volumes of data, waiting to be released. The potential power of this 'big data' does not erase the need for strategic vision or human context and insight.

As data tools and ideas about data usage spread, they will change long-standing notions about the value of experience, the nature of expertise and the practice of management. Smart leaders in every sector already see using data for what it is: *a management revolution*. However, as with any other significant change in business, the challenges of becoming a data-fluent organisation can be enormous and require the right balance between hands-on and hands-off leadership.

Equipping an organisation's ability to manage the increased complexity required of the data infrastructure that will support such ongoing knowledge, data teams (strategists, programme managers, data scientists, engineers, operations and so on) must become an *intentional*, reliable and deeply integrated part of the resourcing strategy in every organisation. Functional leaders in organisations must learn to see data management workers as a regular and deeply integrated part of how they do business in every sector – a central part of every organisational chart and business strategy.

> Leaders must understand that data maturity and nurturing data as an asset for the organisation is never one-and-done, but an *ongoing management and leadership responsibility*.

Every organisation (not just the data team) must become data-focused. We must learn to cultivate a data culture by collaborating more effectively across functions and be able to scope, secure and manage data as a corporate asset. This newly shared understanding increases the supply of data management expertise available for the future. It guarantees a deeper, more inclusive conversation in the data community about how organisations and their customers should manage and use data.

The result: data strategy becomes an intentional component of every business strategy. The by-product of a deliberate data strategy is an increased awareness of data in every function. The dividend of awareness is profound: greater understanding sparks demand for more knowledge and skill development; skilled employees gain an edge in their jobs; organisations with engaged employees cultivate a renewable corporate asset actively; and markets move more quickly to solve enigmatic problems with new velocity.

Organisations are embracing the goal of building strong data cultures. They want employees to gain data skills as the norm, across all sectors. The Bureau of Labor Statistics projects 11 per cent job growth in computer and information technology occupations over the next ten years – much faster than the average for all occupations.[5] Demand for business intelligence (BI) analysts will grow by 21 per cent between 2014 and 2024. A brisk marketplace exists for data management, intelligence and governance services. Companies such as Tableau are setting bold public goals, like their 2021 commitment to enable 10 million data learners over the next five years.[6]

Given the large-scale impact of the role of data in our society, the hard part hasn't been convincing executive leaders that data is important. The core challenge has been getting executives to understand that consuming regular, reliable data is a service that must be invested in – continuously – like a utility. Data teams and business partners must move from ad hoc habits to data intentional, interdependent, strategic alliances. Traditionally, this is where the understanding experiences friction, frustration or confusion.

Today, data leaders need to monitor the constant flow of data like a natural resource. The metaphor of oil has become popular, given its continuous refining process into a new material. But even that metaphor falls short because data requires more than just financial investment. It requires a mindset shift to making data a generalised workforce development skill so the corporate asset is relevant and useful to *everyone* in the company's ecosystem (internally and externally). The cliché 'you get what you pay for' applies here. Data is the gift that keeps on taking as demand for more business and customer intelligence grows.

War stories from executives who have dealt with this dilemma are painful – from leaders who constantly requested data from teams that never produced anything, to data workers who may not have had the skills to deliver despite their best intentions, to organisations who took on low-priority data management projects but then didn't invest in sustaining them to succeed.

Progress towards data-oriented goals is painfully slow for most organisations. Leading corporations struggle to become data-driven. An alarming finding of NewVantage Partners' 2019 Big Data and AI Executive Survey found that:[7]

- 72 per cent of survey participants report that they have yet to forge a data culture.
- 69 per cent report that they have not created a data-driven organisation.
- 53 per cent state that they do not treat data as a business asset.
- 52 per cent admit that they are not competing on data and analytics.

No data leader wants to engage in an initiative likely to fail. No data professional wants to dedicate their career to an effort with a poor success rate. Yet, according to 2015 data, Gartner predicted that, 'through 2017, 60 percent of big data projects will fail to go beyond piloting and experimentation and will be abandoned'.[8] Furthermore, in 2019, Gartner predicted that 'through 2022, only 20% of analytic insights will deliver business outcomes'.[9]

So why do it? Because the need for it will never go away, and it *is* possible. Like any leadership endeavour, data transformation is an art and a science. It takes practice.

---

5   https://www.bls.gov/ooh/computer-and-information-technology/home.htm

6   https://www.tableau.com/about/press-releases/2021/tableau-pledges-train-10-million-data-people

7   https://www.newvantage.com/thoughtleadership

8   Gartner, Gartner Says Business Intelligence and Analytics Leaders Must Focus on Mindsets and Culture to Kick Start Advanced Analytics, Laurence Goasduff, 15 September 2015. https://www.gartner.com/en/newsroom/press-releases/2015-09-15-gartner-says-business-intelligence-and-analytics-leaders-must-focus-on-mindsets-and-culture-to-kick-start-advanced-analytics. This Gartner report is archived and is included for historical context only.

9   Gartner, Predicts 2019: Analytics and BI Strategy, Gareth Herschel, Erick Brethenoux, Carlie Idoine, Austin Kronz, Eric Hunter, Mark Horvath, 27 December 2018. This Gartner report is archived and is included for historical context only.

There will be mistakes and learning along the way. When it works, it becomes the engine behind your business and significantly adds to the value of your organisation.

When leaders start considering data transformation efforts, they are doing so at a time when their teams are deeply siloed and overwhelmed by demand for data services within their function. When the chance to think cross-enterprise is presented, most leaders shrink from the opportunity. 'It's not the way we do things; it's not the [Your Company Here] way!' They have gotten by so far by relying on favours and heroics, but things are just not going fast enough. There are too many errors. The pain is becoming too great. It's like everyone is digging their own well, and many are running dry. Employees burn out, creativity starts to plummet and collaboration points begin to fray.

Data is an asset

It is time to frame the data challenge to the business: *start thinking like a paying customer of a commonly shared asset.* Assets are, by definition, useful and designed for ongoing use. A *data asset* provides the organisation's current, future or potential business value. Across the world, we pay some form of tax for the most efficient management of natural resources or commonly used services and our portion of use. The same can be true of how data organisations realise their funding. Business stakeholders who use data to execute their strategies share the burden of the costs of *utilising* the data. Still, they also contribute to the *asset's value* by monetising that information.

When many data transformation efforts begin as ad hoc habits, there is no practice of taking just enough time to understand what it means to run data as an ongoing service (like a utility). Instead, we launch-and-leave reports and store-and-ignore our data, leaving interpretation, context, security and customer trust to chance. We rationalise these less mature behaviours as 'getting it done' or 'being scrappy'. While true at the beginning, these behaviours become habits and end up costing more down the line. Ad hoc efforts do not deliberately cultivate data as an asset.

Not all data transformation efforts run poorly. However, in many cases, data leaders do not realise when they are in the best position to proactively adjust their approach to seeding the data revolution. They do not seize the opportunity to influence their stakeholders' behaviour and expectations. Making data transformation efforts as reliable as running a utility or stewarding raw resources means starting with the needs of the business stakeholders rather than the available resources in the data transformation team as they currently exist. It means creating a clear scope of work and a project plan rather than 'being scrappy', 'getting it done' or 'winging it'. We can still be scrappy while being intentional about our project, moving from running a utility to curating a valuable data asset.

This guide includes exercises, templates and worksheets to guide the main aspects of a data project. These materials supply a method of battle-tested best practices for driving a series of data projects towards enterprise-wide initiatives that integrate and support business strategy.

Those looking for more information can follow the research footnotes throughout this book, highlighting barriers and opportunities for organisations considering data transformation efforts. The Data Management Association (DAMA), DataVersity, BCS The Chartered Institute for IT (BCS) and the EDM Council share additional best practices for organisations that truly embrace data transformation efforts. The Project Management Institute (PMI) and ProSci Change Management organisations supply many free resources cited throughout the guide and in the Professional Resources section at the back of the book.

Developing a learning culture through digital upskilling is not a separate deliverable from the data team's need for a data-forward culture through increased data skills. This shared goal positions human resources as one of the most strategic (potential) partners of data transformation efforts. While it depends on the organisation's structure, the HR department designs culture as part of its mandate. The mechanisms they use – such as the organisation's values, job competencies, performance measures, leadership development programmes and coaching tools – complement effective driving of data projects. They facilitate senior leadership retreats, design leadership surveys and recruit talent that drives data cultures forward. Each HR programme is a potential point of both partnership and intervention for landing data-forward behaviours, providing case studies and emphasising data-forward language. Descriptive behaviours help paint the picture of what we want to step towards.

While measuring and evaluating the value of data transformations is hard (especially in the early phases), approximately 25 per cent of an organisation's budget is dependent on data processes and services supporting their efforts.[10] Determining data cost can vary across industries. For example, the consumer-packaged goods sector doesn't directly engage customers and has a higher relative spend on data sourcing. The results are the same: managing data is a significant cost for most organisations.

Capability and confidence for understanding and driving data projects won't develop organically – they must be actively nurtured. We must apply direct interventions, programmes and systemic ways to foster data awareness and have an effective, inclusive and intentional method for driving data projects in organisations. One way to do this is by having a rigorous and inclusive approach to driving data projects.

But first, let's learn about data as a concept and its journey to become an asset.

WHAT IS DATA?

Merriam-Webster defines data as:

1. NOUN
- facts and statistics collected together for reference or analysis;
- the quantities, characters, or symbols on which operations are performed by a computer, being stored and transmitted in the form of electrical signals and recorded on magnetic, optical, or mechanical recording media.

2. PHILOSOPHY
- things known or assumed as facts, making the basis of reasoning or calculation.

---

10 'Reducing data costs without jeopardizing growth', July 2020, McKinsey & Company. https://www.mckinsey.com/capabilities/mckinsey-digital/our-insights/reducing-data-costs-without-jeopardizing-growth

That data is both a noun and a philosophy is significant. It is *both* art *and* science. Philosophy is a rational investigation of the truths and principles of being, knowledge or conduct. Ultimately, philosophy uses data in a search for wisdom. Data as facts and information isn't wisdom; it must first go on a journey and experience transformation. See Figure 1.3.

Figure 1.3 Cartoon by David Somerville, based on a two-pane version by Hugh McLeod (Source: © Gapingviod Culture Design Group LLC 2023)

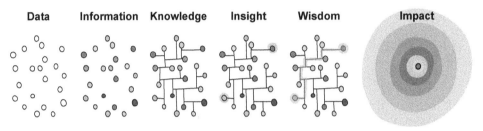

In this way, data is a kind of medium for expression. It is information, full of emotions, a language, an art, and an incredibly valuable asset if managed well. Data is an alchemist's dream – it changes, increasing or decreasing in value, as it moves through a supply chain (which we'll get to later).

Let me explain.

Data as information

Data refers to discrete pieces of information, usually formatted and stored in a particular way for a particular purpose. Data needs to be 'structured' for a specific purpose to create value and be effective. As such, it cannot always serve multiple objectives if it is not structured to do so. Data can exist in various forms. It can be text on paper, code stored in electronic memory or information passed from person to person. Today, data most commonly refers to information transmitted or stored electronically.

All data can be categorised as machine-readable, human-readable or both. Human-readable data utilises natural language formats (such as a text file containing ASCII codes or PDF documents), whereas machine-readable data uses formally structured computer languages (Parquet, Avro and so on) to be read by computer systems or software. Some data is readable by machines and humans, as in the case of CSV, HTML or JSON formats.[11]

The line between machine- and human-readable data is becoming increasingly blurred because many formats today are accessible enough to be navigated by a human yet structured enough to be processed by a machine. This is primarily the result of artificial intelligence, machine learning and automation, which streamlines tasks and workflows so manual data entry and analysis is done by a machine rather than a human. However,

---

11 https://www.webopedia.com/definitions/data/

these processes must maintain human readability if the programming needs to be adjusted. Most data in these cases also exists in a vacuum and does not have much meaning without context from a human perspective.[12]

Data prompts feelings

We might feel confused or intimidated when seeing data or when asked to justify our efforts. When confronted by calculations we do not understand, many feel painfully embarrassing maths-shame, recalling early years wrestling with algebra and calculus. Others feel curious and view data as an intellectual puzzle to be solved, anticipating what might come next. Some experience both feelings at once! We experience a broad spectrum of emotions, such as anger, delight, fear, excitement, disappointment, surprise, confusion, joy and satisfaction, when reviewing customer feedback, personal health data, monthly business reviews, retirement projections, weekly flash reports, daily manager ratings, bank statements and annual performance reviews.

We are awash with information in our lives and work. Wherever we look, there is a constant flood of facts, statistics, models, visualisations and projections to process, examine and make meaning of. Our minds have become numb to the constant stream of data generated by our lives and work.

Yet, there is a strange desire for *more* – almost an *addiction*. If we are provided monthly data, we eventually want up-to-the-minute data. If we have global data, we want it by country – sometimes even by zip code. We want the latest, most granular data possible. We become irritable when data is inaccurate or unavailable and costs us more than expected. Too much data and we risk the analysis paralysis that accompanies cognitive overload.

Eventually, we learn that data must be focused and used responsibly. We must invest the emotional and intellectual energy necessary to synthesise and interpret these numbers into words, maybe pictures. We must overcome the inevitable insensitivity which we develop to what the data might mean or instruct us to do.

At a time of unprecedented access to information, we become frustrated by the lack of action to solve our organisation's (and the world's) problems. Many of these problems are systemic and deeply rooted in seemingly intractable cultural patterns and routines. To sense the significance of these issues and begin to confront them, we become aware of our (and others') *feelings* around data that allow and inhibit our ability to work with data effectively.

Data as language

Data intelligence is based on the language around data.[13] Considering data as a second language might elicit memories of what learning high school French or Spanish felt like or what it was like to grow up bilingual. Thinking about data in the context of literacy makes sense because we can add fluency and mastery skills, but if we start with fluency, we will leave a lot of people out. Language permeates our lives and cultures. It connects

---

12 https://regeneguwutow.weebly.com/uploads/1/4/1/3/141391647/taverazubi_fenir.pdf

13 Valerie Logan (former Gartner analyst), CEO of The Data Lodge, codified the definition and foundation for data literacy research and focuses on the pioneers of data literacy programmes.

or isolates us as individuals, organisations and societies. We consume data about our health, finances, the products we consume and the news. At work, we test hypotheses, check assumptions, monitor progress to achieve results and make data-based decisions. All these examples make using data as a language more personal and grounded in the reality of our lives, rather than thinking of it as only something a statistician does.

Why does language matter?

Not everyone will become data-fluent but everyone must gain a level of basic literacy. As the data literacy space continues to mature and establish standards, it is essential to remember that every data project presents an opportunity to think beyond the Three Ps. Every data project is an opportunity to revolutionise your organisation's culture to become more data-forward. Data provisioning by data teams must be met by stakeholders who can speak, write and use data *in context*. Therefore, providing and consuming data must be intentional and strategic to contribute to data as a corporate asset.

Data as art

Nothing communicates a message quicker than a good (or bad) chart. The *art* component of data visualised – or graphical excellence – is the efficient communication of complex quantitative ideas to an audience.[14] This requires clarity, precision and efficiency.

Compelling visualisations combine disciplines by joining data with story. Visualising data is integral to analysts' and data scientists' daily lives. Communicating impact using data is quickly becoming part of everyone's job – across all disciplines.

Good data visualisation requires clean data that is well-sourced and complete. A visualisation's sole goal is to make processing information easier for the human brain and, therefore, for the audience.

A by-product of effective visualisation is better decision-making (actionable insights), meaningful storytelling and increased data intelligence (for example, data literacy). There are many creative examples (with tutorials) of complex ideas being communicated with clarity, precision and efficiency. Newspapers, periodicals and consulting reports have great examples of visualisations simplifying complex concepts. Some of these are truly art.

Data as an asset

There is much debate about when data becomes information and, from there, when information becomes insights. This guarantees equal discussion on what 'data as an asset' means.

For this guide, data is a corporate asset when it provides *accurate*, *current* and *continuous* insight into a business's operating model (for example, performance, operational efficiencies and strategic direction). Physical corporate assets include databases,

---

14 Edward R. Tufte, *The Visual Display of Quantitative Information* (Graphics Press 2001). The viewer should be given the greatest number of ideas in the shortest time with the least ink in the smallest space: the 'less is more' philosophy in the context of data visualisation.

data files, contracts and agreements, system documentation, user manuals, training materials, operational/support procedures, business continuity plans, backup plans, audit trails and archived information.

Data is only an asset when it is understood and correlated effectively to identify underlying business problems and when it contributes to a solution to address them. Data is *not* valuable simply because it is data. The challenge for any organisation is to get tangible benefits from the data (for example, financial, retail and sales data), which can only happen if there is a high-performance team of data analysts, statisticians and data scientists in place to draw insights.

In his book *Data Driven: Profiting from Your Most Important Business Asset*, Tom Redman suggests that for data and information to become assets, three simple actions are required. An organisation must:

1. provide for the 'care and feeding' of its data and information, just as it does for its capital and assets;

2. put its (high-quality) data and information to work in unique and interesting ways; and

3. fully embrace the potential of data by learning to make better, more confident and more informed decisions, tune into the growing needs for data and information and accelerate work to bring them to market.

If a data team does not have representation during business strategy discussions, it cannot effectively estimate the resources, tools and platform costs to support it. Efforts are likely to be chaotic and undisciplined.

Data enables humility

Data is all these things and more. Data is often regarded as driving stronger decision-making. It is used as a catalyst for curiosity. All data efforts support enabling everyone to be more confident and comfortable with decision-making.

Data should also be a source of great humility for leaders at every level. Everyone needs to be more comfortable and confident communicators with data. We must challenge where the data came from and understand if it is clean (enough). Observing a data trend is sometimes sufficient to make a solid decision. For other decisions, specific counts are needed. Conversations about data, such as understanding our colleagues' logic in building a business case or justification of a decision, are just as valuable as the data itself.

Humility must be modelled. Leaders must demonstrate genuine humility with data, especially senior sponsors; they are some of the most vital teachers in the organisation. Managers and leaders carry culture through their behaviours. People around them spend considerable time thinking about how and why leaders make specific comments or decisions. How everyone regards or doesn't regard those behaviours is critical to the success of a data project. When a leader takes time in a meeting to ask how a term is defined or what 'AI Ethics' means in a particular use case, they provide critical context for others on how to model, structure, store, share and otherwise use data. When leaders demonstrate they don't know something, or seek clarity, they model the humility necessary to use data responsibly.

WHY DATA TERMS MATTER

Terminology on data projects can get murky fast. Business insights and the underlying data architecture to support them have existed since the first computers; yet, every 20 years or so, the data industry goes through a shift in vocabulary. We are in the midst of that now.

The search for the perfect 'business insight system' has had many popular terms. What we refer to more comprehensively now as the 'data supply chain' started as Executive information systems (EIS) and decision support systems (DSS) in the 1980s. In the 1990s, data warehousing (DW) and business intelligence (BI) became more popular terms. Performance management (not the kind in human resources) had some attention in the early 2000s, quickly giving way to predictive data, artificial intelligence and automation efforts. Most roles associated with these efforts were variations of the 'analyst' title.

The commoditisation of cloud data storage solutions and cloud computing shifted the focus from data aggregation to data extraction of useful information from data at scale, creating new opportunities for higher specialisations. Therefore, new roles and titles appeared, such as 'data scientist', and the term 'data science' started appearing in job descriptions for many roles: 'data science, analyst', 'data science, research scientist', 'data science, applied scientist'. When the term 'data science' started to become too vague, more specialised titles began to appear, such as 'machine learning (ML) engineer'.

Mattia Ciollaro and Fabrizio Lecci illustrate the overlap of skills across different roles. Such overlap can have advantages and disadvantages. On the one hand, it can generate confusion regarding the division of responsibilities and career development. On the other hand, it allows a data team to gradually mature without having to immediately hire all the different roles: in addition to their core responsibilities, data engineers can set up basic metrics and dashboards; data analysts can run some regression models and A/B experiments; data scientists can partner with software engineers to deploy their models. Skill overlaps allow teams to evaluate the need for additional layers of complexity without necessarily pausing their work and waiting for more resources. Some teams go as far as establishing full-stack data roles, requiring experience from data collection and storage to data analysis and visualisation. Individuals in full-stack data roles are responsible for handling all aspects of data management, from the raw data sources to making data-driven insights accessible to stakeholders. These roles are often multidisciplinary and require a broad skill set that spans various data-related technologies and tools. However, most companies differentiate roles and titles when the need for specialisation becomes evident. The graph in Figure 1.4 intuitively shows how much time each role typically spends on different tasks and the overlap of skills. This can vary significantly across companies, especially when only a subset of these roles exists in a team.

Terms influence roles

There is a saying: 'Language is a window to the soul.' Similarly, the terms we use to reflect our job titles and tasks are a window to our data cultures and our technical evolution – in real time!

Consider the technological changes in code generation just in our lifetimes. Early on, engineers wrote machine or assembly code, low-level programming languages that communicated directly with a computer's hardware. They then learned high-level

Figure 1.4 Skills overlap on data teams (Source: Shown with permission: https://www.datacaptains.com/blog/guide-to-data-roles)

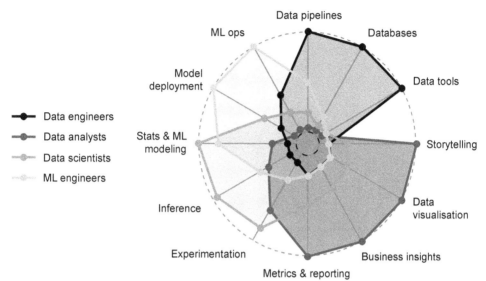

computing languages (Python, Visual Basic, Java, C++) that could be executed directly on a computer. From there, rendering what was happening on a browser locally was linked to what was happening in a cloud. Code became object-oriented. Objects had code *and* data associated together. That led to a world with Javascript objects on a local browser being executed to cloud objects on a server somewhere else, talking to a database to make things happen. Product managers in the 1990s created specifications, determined requirements and selected user stories to define product requirements for an engineer to execute source code and objects.

All that has happened in the pre-Web, Web 1.0 and Web 2.0 worlds.

Today, we are moving towards ChatGPT, a large language model exceptionally efficient at generating structured text (that is, *code*). We can give it a simple prompt, 'Build me an algorithm' or 'Build me [something] using [any] language code that does [something]', and it will execute something that works [mostly].

What are the implications?
The next generation of coders will completely rely on these generative computing tools. Many engineering and coding roles will become more like the roles of product managers in the 1990s. They will create specifications, determine requirements and select user stories. Instead of these being given to engineers (coders), they will feed these requirements to code generators that Microsoft and other companies will build into their coding tools, and those products will generate code that attempts to accomplish that user story or task.

Engineers have shifted entirely from writing code at the lowest level to developing input requirements (specifications, requirements and user stories) for an AI-code generator based on a large language model, which will create something that works.

It is still incumbent on the data consumer (generally a data team and their stakeholders) to know how data works. We can't let any data solution turn into black boxes. Whenever we receive output from AI or ML, we must have the skills to challenge it. We must consider such output as 'the first pancake', understanding exactly how it is flawed and how to challenge the ingredients, the preparation and so on, to defend our logic with an answer that aligns with 'a better pancake'.

This book intentionally opts to use the term 'data projects' for anything related to data transformation. There is incredible flux and confusion right now as organisations define for themselves the distinctions between project, programme and product management or how they delineate between business intelligence and data science. It doesn't help when data products, services and consultants co-opt these terms as value statements or features.

Looking at the popularity of two buzzwords over time – 'product management' and 'programme management' – it's clear that there is a recent spike in interest.[15]

Did companies not do data science before 2011? Yes, but everybody used a variation of the 'analyst' title. For example, until 2015, most analytics and data science roles at Google tended to fall into the 'quantitative analyst' bucket. With the commoditisation of cloud data storage solutions and cloud computing, the focus shifted from accumulating data to extracting useful information from data at scale.

So, what should be the composition of a data team? What roles are needed for what tasks? It can be confusing. Let's take a step back before we try to answer these questions.

THE DATA JOURNEY, FROM INFORMATION TO ASSET

'If you can't describe what you are doing as a process, you don't know what you are doing.'

– W. E. Deming

Increased awareness and skills make us better data advocates for appropriate data usage and ongoing data practices. It is here we begin to touch data quality. Data quality is not simply about getting the 'facts right' inside databases; it is about developing a reasonably complete and accurate picture of current state on problems of broad importance. We will know when to constructively speak up if standards are inadvertently missed or knowingly bypassed and how to use data more ably to influence others. Without that context, data is hard to curate as a corporate asset, and eventually monetise. Without this awareness, we won't know if monetising it is ethical or not.

Consider how consumer data is used and traded today. Many technology organisations are inclined to review their privacy and security governance policies when popular companies find themselves in the headlines. Zoom settled $85 million in claims after ensuring end-to-end encryption to protect users' data while sharing that same data with

---

15 Google Plot: https://trends.google.com/trends/explore?date=all&geo=US&q=product%20management,program%20 management

Facebook and Google.[16] Apple looks for abusive content by scanning 'private' photos on people's iPhones.[17] While organisations defending civil liberties point out policy inconsistencies and the essential roles such companies have in protecting consumer privacy and security, such headlines motivate the public to collect petitions and write open letters to the leaders themselves.[18] [19]

Every level of the organisation – from top management to line level – makes decisions regarding data sharing, storage and disposal. As the Zoom and Apple examples indicate, there is no organisation that, after an internal or external audit, would not find areas for significant improvement of their data practices. While the intention of flagging malicious and illegal content to law enforcement is positive, the precedent should be a wake-up call for data leaders. A company acting on its own or at the request of a government can look through a customer's personal data, without their knowledge or consent, using an opaque algorithm, and take actions against which the customer has little or no recourse. Data teams and business stakeholders must collaboratively navigate these ambiguous challenges and high-stakes trade-offs.

How data (and IT) are not just utilities: they create assets

Most of the time, the focus for data projects is on incrementally improving existing 'analogue' processes or products. We want to help things run faster, better or cheaper. Used in this way, data operates as a utility. It optimises the sales, production and support pipelines of an organisation. It generally sits and is managed in the IT discipline, which is also (often narrowly) viewed as a utility.

When you turn on the tap, you expect water to flow. When you plug a device into the wall, you expect electrical power. When a business provides data requirements, it expects reports, answers, and insights, in just the same way. We also often imagine data as something collected and stored for future use, like water in a tank or energy off a grid – or a filing cabinet full of records that we'll return to later.

Why isn't data like a utility in most organisations? Why is cultivating data as an asset so hard?

Part of the issue concerns the complexity of data and the near-infinite combinations required to support the business. The planning and engineering involved with ensuring the reliable and safe delivery of water and electrical services can be complex. However, once the infrastructure is in place, the utility delivers the product without significant variation from customer to customer.

Most organisations would not dream of building their own water and electricity infrastructure. Yet many organisational business units readily accept responsibility for

---

16 https://arstechnica.com/tech-policy/2021/08/zoom-to-pay-85m-for-lying-about-encryption-and-sending-data-to-facebook-and-google/

17 https://www.nytimes.com/2021/08/05/technology/apple-iphones-privacy.html

18 The EFF is the leading nonprofit organisation defending civil liberties in the digital world and champions user privacy, free expression and innovation through impact litigation, policy analysis, grassroots activism and technology development. Its mission is to ensure that technology supports freedom, justice and innovation for all people of the world.

19 https://www.eff.org/deeplinks/2021/08/apples-plan-think-different-about-encryption-opens-backdoor-your-private-life implementing independent data efforts. They assume this is the best way to get the data customised to the needs of their business and available budget. Having so many siloed efforts, with their unique perspectives on standard measures, makes curating a corporate asset among these data sources nearly impossible.

The reliability of a data infrastructure is also heavily influenced by budget and available expertise. With water and electric utilities, the provider maintains the infrastructure. The utility company conducts the repairs if there is a break in the water main or downed power lines. With ad hoc data efforts spread around the organisation, various business units devote significant resources to maintaining independent (rogue) data stacks and responding to issues when they arise, distracting them from core business functions. There may be good reasons for this approach. However, when managed in this way data is not a streamlined utility, not strategic and not on track to becoming a corporate asset.

If we want to increase our maturity and fluency with data to take advantage of data's full potential, we must start thinking of all things digital and data as things that expand the limits of what is possible. Data is increasingly central to value creation, exponential business models and innovative technologies. We need to be able to access, combine and transform the data we have quickly – more akin to a plumbing system than a filing cabinet. To work effectively, we must build basic awareness of the data supply chain and basic data literacy skills in every part of our organisations.

It's estimated that the world produced 79 zettabytes of data in 2021.[20] That means that globally, we aren't struggling to create data; we're struggling to help leaders use it effectively.

$$79,000,000,000,000,000,000,000$$
(that's 21 zeros!)

Research shows that 70 per cent of complex, large-scale change programmes don't reach their goals.[21] Years of research on transformations shows the success rate for these efforts is consistently low: less than 30 per cent succeed.[22] Failure to reach goals occurs for many reasons – inconsistent buy-in, lack of leadership urgency and/ or investment, lagging technologies and teamwork falling apart, to name a few. For any digital transformation to sustain momentum, employees must have digital fluency, including an understanding of the building blocks of digital, starting with data.

Data has a supply chain – from information to insights. This supply chain has two main components: human and machine. On the human side, we need to gain more awareness of how we interfere with data today to develop more objective, thoughtful interventions. On the machine side, we need to understand the layers of systems through which data

---

20 https://firstsiteguide.com/big-data-stats/#:~:text=In%202021%2C%20the%20overall%20amount,amount%20is%20 expected%20to%20double

21 'The "how" of transformation', May 2016, McKinsey & Company, www.mckinsey.com. © 2023 McKinsey & Company.

22 'Unlocking success in digital transformations', October 2018, McKinsey & Company, www.mckinsey.com. © 2023 McKinsey & Company.

travels and how it changes from data to information to insights that inform and influence our behaviours, our decisions and how we operate.

The human side

The irony with data projects is that to gain greater mastery of the technical aspects of data, we need to learn less about technology and more about how to *think*. Thinking is an innately human – philosophical – process.

Our current knowledge and technology only work when the problem definition, solution and implementation are clear. Most data projects are the *output* of a known process, yet the insights they generate help stakeholders generate solutions when the problem definition, solution and implementation are *not* as clear. Thoughtful insights depend on high-quality data and kick off creative problem-solving activities to resolve core business challenges.

However, these efforts leave a lot on the table. We forget that *current thinking* organises *current technology*, which generates output in line with *current expectations*. We treat surprises as issues to be resolved. This dynamic creates a strong pull to see or bring results 'back in line' with expectations, rather than a jumping-off point for alternatives or opportunities.

Humans make a significant contribution (both positive and negative) to creating a data asset that is often under-scrutinised. How we intentionally intervene and unwittingly interfere with data as it moves through various systems is often taken for granted to such a degree that we avoid accountability for our actions altogether.

To become the best partner to the machines we rely upon, we must learn to embrace a few thinking principles that will stretch us from our current comfort zones. Specifically, we need to:

- challenge existing paradigms;
- computational thinking;
- increase our dynamic, ongoing learning;
- find our learning edge.

Challenge existing paradigms

Paradigms are a framework for 'concepts, values, perceptions, and practices' that help us understand our world, ourselves and our work.[23] Paradigms cover many aspects of our lives, such as what kind of car to buy or what type of career to have. However, one of the most important paradigms is our worldview, a set of constructed perceptions and ideas about how we think the world works. Governments, organisations and our own families instruct us on what paradigms to adopt as we develop. They do this through incentives and disincentives until the desired behaviour occurs. Over time, we acquire knowledge, processes and skills supporting a certain standard. All this creates culture, as illustrated in Figure 1.5.

---

23 F. Capra, *The Web of Life: A New Scientific Understanding of Living Systems* (Anchor Books 1996).

Figure 1.5 The relationship between paradigms, mental models, mindset, behaviours and culture (Source: https://www.semanticscholar.org/paper/The-psychometric-properties-of-a-talent-mindset-Welby-Cooke/7de597f67cd51a31f290a6a9a3b134ab56ace0be)

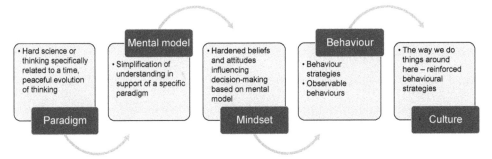

Shane Parrish and Rhiannon Beaubien authored a series of books called *The Great Mental Models*. Their three-book series covers *General Thinking Concepts* (vol. 1), *Physics, Chemistry & Biology* (vol. 2) and *Systems & Mathematics* (vol. 3). The series centralises and summarises popular concepts for improved decision-making. I recommend them for their accessibility as a canon to refer to for details of what you already know (or should know from school).

Mental models help us simplify complexity to understand the world more easily. They shape what we think and how we know. They influence and shape the connections and opportunities that we see. Mental models help us determine the relevance of some things over others, and how we reason. For example, referring to data as a 'utility' or 'natural resource' (like oil) is shorthand for trying to understand the complexity of data and how it is managed. Mental models that remain unchanged or unchallenged for too long become mindsets that influence decision-making based on behaviour.

The quality of our thinking is largely influenced by paradigms that influence our behaviours. So, what if that way of framing data is blocking our progress? We need accurate models in our minds, certainly. But we also need to establish new models to acknowledge that the data might be more complex than a utility or refining oil.

We have all experienced this first-hand. Engineers think in terms of systems. Psychologists think in terms of incentives. Businesspeople think in terms of cost and risk–reward. Everyone looks through their unique 'paradigm glasses' fashioned by their life experience, culture and job function. Trying different glasses requires as much unlearning as it does learning.

> 'The chief enemy of good decisions is a lack of perspectives on a problem.'
> – Alain de Botton[24]

It is essential to know how to shift our thinking before we ask others to shift core beliefs, alter their habits or change their minds. First, we need to understand the steps required to change thinking.

---

24 Alain de Botton, 'How to Decide', The School of Life. https://www.theschooloflife.com/article/how-to-decide/

Step 1. Recognise when paradigms need to shift

Depending on the same paradigms, with a fixed mindset, for too long offers diminishing returns and can put us in a rut. Beliefs and behaviours must shift to gain new results and maintain progress. When caught between two competing objectives (global versus local, hierarchical versus flat and so on), the answer lies in the centre of the polarity. For example, companies today want to be local *and* global, grow organically with individual customers, *and* become enterprise leaders. Managing trade-offs like these demands a new way of thinking, from either/or (old way or new way) to both/and. See Table 1.4.

Table 1.4 Applied practice: learn to shift paradigms

Consider your current organisational challenge.	
Observe areas where old ways of thinking are no longer working.	
What's a familiar business habit, dynamic or practice that no longer works?	How is the landscape changing in your sector? Where would new ways of thinking help reframe existing problems?
We measure our success based on individual customers; our growth has been 100% organic – but we need to scale faster than individual customers can take us.	*With a recent acquisition, we bundle ourselves as an enterprise solution but don't know how to position ourselves as a leader to individual customers and the enterprise.*
Now you try:	

Step 2. You are here, but where are you going?

The process of unlearning feels more challenging when the next step isn't clear. The gap between 'here', where current thinking is familiar, and 'there' (the unknown) seems broad. We can feel vulnerable to the risks. We only need *just enough* clarity to experiment in a new direction. Even a tiny pivot can broaden our field of vision and sense of opportunity.

From this new vantage point, we can see how previous ways of thinking and familiar tools might still be helpful or might be getting in our way. If we start to play with them – to break them down or combine them in new ways – we can form new ways of doing things and of thinking, and potentially new solutions. These pivots in thinking are how we can actively change *along* with our environment.

Step 3. Provide a path towards a better future – then take the next step

This step is the hardest part. When leaders invest in innovation or a new strategy, they are doing so because they see diminishing returns from status quo thinking. Moving towards a new vision is no guarantee of success. Most people executing the status quo don't see a need to shift and want to see a clear picture and all the risks removed before taking that step, making change management very hard.

However, everyone needs to find a path through for themselves (meaning: everyone's thinking needs to change to some degree) so the benefits become more apparent. Many challenges stand in the way of progressing new ideas. Believing they will happen because it's the right thing to do is the fastest way to undermine their progress. That is one of the reasons why shifting culture from ad hoc efforts to operationalising data is so hard and takes so much persuasion and preparation. See Table 1.5.

Table 1.5 Applied practice: shifting paradigms

Practise how to change your thinking before trying to change others'.

Determine whose thinking and behaviours need to shift; this is typically called the 'change target'. Complete these steps for each stakeholder group of people you are trying to influence. Try identifying with the target group by using 'we'.

1. Identify the domain you're transforming. For example, data is primarily reports and dashboards.

2. Substitute the words in the [brackets] with your scenario: Today, we consider [a domain] as [prevailing paradigm]. Example: Today, we think about data as visualisations.

3. Find the words to reframe that paradigm: In the future, we could consider [a domain] as [new paradigm]. Example: In the future, we could view data as a common language between disciplines that will change the social circuitry of organisations.

Now you try:

TODAY	Group	FUTURE
Today, we consider...		We could consider...
_____ (domain)	_____ → → →	_____ (domain)
as _____ (current thinking, behaviours)		as _____ (new thinking, behaviours)

Computational thinking

Working with data requires breaking down problems into distinct parts, looking for similarities, identifying the relevant information and opportunities for simplification and creating a solution plan.[25] There are four stages of computational thinking that help drive data projects effectively:

- Decomposition: break down something large into several smaller components.
- Abstraction: develop a model or representation simplifying a key concept.
- Patterns: use reusable components to minimise error and work.
- Algorithms: a series of rules for solving a particular kind of problem.

Computational thinking drives everything we do – from planning a trip to leading an initiative. We have been conditioned for decades to think like machines. Computational thinking is the most common approach to problem-solving, dominates our educational curriculum and is the foundation of problem-solving in organisations. Anyone who has ever simplified a problem or improved a process has engaged in this thinking. More than marketing or buzzwords – *applying* these terms in our day-to-day life helps us to think more deeply and discuss technology meaningfully.

Computational thinking is an area in which graduate students and those new to data management have a hard time, because they don't know what they don't know. They might realise they need to interview stakeholders about pain points, but they might not know all the tasks contributing to the deliverable. The key is thinking through to the end of the deliverable. It may surprise you that most of the upfront effort for driving data projects goes into project scoping and planning, where each deliverable breaks down into elemental tasks. Assigning deadlines and dependencies for these tasks is critical for driving accountability and enabling stakeholders to take ownership of their responsibilities. Sound like common sense? It is, but only when project managers and leaders admit they can't keep track of it in their heads! See Table 1.6.

Table 1.6 Applied practice: think like a machine, part I

Break down problems. Consider a typical data project; break it down into smaller tasks.		
Process	**Key steps**	**What do you observe?**
Run monthly report	*Gather business requirements*	*Stakeholders often do not supply context; it is difficult to get complete requirements.*
	Identify data sources	*Data is unreliably delivered. Need to set up a service-level agreement for data owners to provide data regularly.*

(Continued)

---

25 The Tech Interactive: https://www.thetech.org/sites/default/files/techtip_computationalthinking_v3.pdf

Table 1.6 (Continued)

Process	Key steps	What do you observe?
	Compile report	*Some parts of data delivery can be automated.*
	Check the report for accuracy	*Opportunity to analyse reports for insights. Opportunity to revisit stakeholder roles and responsibilities and training opportunities.*
	Prepare communication to stakeholders	*Opportunity to collect stakeholder satisfaction on a semi-regular basis. Confirm the business value of the data insights being provided. Solicit ongoing data requirements.*

Now you try. Think of a problem or process that could benefit from being broken down into individual steps. What key steps can you identify? Looking at the list of steps, how does computational thinking change the way you see the problem?

Process	Key steps	What do you observe?

Table 1.7 Applied practice: think like a machine, part II

Operationalise problems into viable work packages. Now, consider how smaller data projects become more significant initiatives. Your problems might align well with a senior leader's commitments. What could be improved, modified or added (regardless of whether it falls within your job description or your stakeholders')?			
Problem to solve (project) *Enable business decision-making across the division*	Task 1 *Develop reporting packages for ongoing delivery*	Subtask 1.1 *Requirements gathering*	Work package 1.1.1 *Meet with all stakeholders*
			Work package 1.1.2 *Determine initial report templates (user experience (UX), cadence, etc.)*
		Subtask 1.2 *Secure data sources & confirm data contracts*	Work package 1.2.1 *Identify data sources*
			Work package 1.2.2 *Identify data owners and secure service level agreements (SLAs)/contracts*
	Task 2 *Support reporting effort via comms*	Subtask 2.1 *Connect report requirements to business insights and impact*	Work package 2.1.1 *Develop an editorial calendar for ongoing communications highlighting the data team's impact on driving business value*
			Work package 2.1.2 *Develop communications templates and training agenda*
Note: the example is not exhaustive.			

(Continued)

Table 1.7 (Continued)

Now you try:			
Problem to solve (project)	Task 1	Subtask 1.1	Work package 1.1.1
			Work package 1.1.2
		Subtask 1.2	Work package 1.2.1
			Work package 1.2.2
	Task 2	Subtask 2.1	Work package 2.1.1
			Work package 2.1.2

Dynamic, ongoing learning

We do not walk into an organisation and see dynamic skills. We observe a particular organisation's dynamic practices, unique way of doing things and collective mindset. These observations can appear in certain behaviours: discussion and learning from failed experiments or fear of failure, open or closed communication and flatter or bureaucratic hierarchies.

Consider how your organisation defines the concept of 'dynamic learning' for itself. How do your leaders avoid long-term failure while focusing on short-term success? How does your company become more innovative (as a proxy for growth)? These ways of working permeate a culture and manifest as a way of doing business – such as the Microsoft Way, the Global Petroleum Way or the Mitsubishi Way.

The academic term for this kind of dynamic learning is *ambidexterity*. Moving past what served us at one time is *not* just about replacing old knowledge with new information. It is about the ability to *switch back and forth dynamically in response* to changing conditions.[26] For example, iterative improvement of the status quo is helpful for ongoing optimisation, balanced scorecarding and cost-reduction activities. However, there are other contexts where more strategic thinking is desirable – creating new business value, solving seemingly intractable problems or launching new communities and markets. The ability to develop and switch between two or more useful models is ambidexterity. People and organisations who can operate between two dynamic thinking states (in response to changing conditions) are resilient and *valuable*.

When business leaders describe their organisational cultures as embracing what it means to 'fail fast' or suggesting that we are all in a 'learning organisation', it rings hollow if the aspirations are not lived and structurally supported. Employees see through the public relations of an organisation that markets itself as innovative yet executes the same playbook year after year.

Table 1.8 illustrates the dichotomy of dynamic learning in the real world.[27] Generally, organisations are hierarchical or flat, closed or open, dogmatic or experimental because they optimise for specific outcomes: scale and productivity or speed and scale.

The speed of business today pressures leaders to drive for *all four* characteristics simultaneously. For that to happen, employees need to become more fluid between what appear to be oppositional states. We must have structure, standards and guidelines to direct us when status quo paradigms fall short. That does not mean that we abandon structure whenever we feel like it. It means we use structure when it makes sense and are empowered to switch to guidelines when it is most constructive.

Table 1.8 Dynamic learning in the real world

	Scale and productivity	Speed and creativity
Steering		
Direction setting	Autocratic and strategic clarity	Iterative collaboration and shared vision
Decision-making	Data and rationale	Conviction and belief
Performance management	People and groups/team performance	Guidelines and values
Transformation		
Mentality		Proactive and evolutionary
Mechanism	Temporary structures and virtual teams	People and embedded routines
Motivation		Personal advancement and inspiration

(Continued)

---

26 https://hbr.org/2016/11/why-the-problem-with-learning-is-unlearning

27 https://www.cutter.com/article/ambidextrous-organizations-how-embrace-disruption-and-create-organizational-advantage-502561

Table 1.8 (Continued)

	Scale and productivity	Speed and creativity
People		
Roles and responsibility	Controlled	Guided
Knowledge management and learning competencies	Technology and expert-based	Trial and collaboration-based
	Professional skills	Self-management and self-development skills
Culture		
Leadership	Transactional and formulated	Empowered and transformational
Corporate climate	Values and relations	Innovation and market Creative and disruptive
Corporate purpose	Shareholder and stakeholder value	Sustainable development
Structure		
Organisation setup	Scale/function-centric	Market/customer-centric
External collaboration	Information gathering	Value creation
Internal collaboration	Hierarchy and cross-functional teams	Collab platform and collab environment
Process		
Process management	Standardised	Adaptive
Process governance	Report-driven	Process/department-driven
Systems and technology	Report and decision supp/ efficiency and automation	Effectiveness and collab insight and foresight

To develop and maintain dynamic skills and become more dynamic, organisations (and people) continuously seek to add capabilities to drive disruptive new business model innovations. Dynamic learning seeks to enhance the organisation's operational experience with emotional connections to new customers. To do this, everyone must set their development goals within a set of target values and behavioural characteristics displayed in *both* columns. Organisations clinging to an either/or culture will reduce their potential impact. Organisations that embrace a 'both/and' approach to thinking and execution will increase their potential impact. See Table 1.9 and use Worksheet 1.

The key takeaway with this essential thinking skill is adopting a mindset which seeks ways to accomplish synergies between the skills you rely on today and the skills you need to develop. The training manual you generate from this skill-building is no longer static; it must constantly adapt to changing conditions.

Table 1.9 Applied practice: dynamic learning

Help balance current and future skills.

Superpower: connector

1. What is your superpower? Why do people ask you for help?
2. What tasks do you avoid? What projects make you feel less competent?
3. How can you develop new skills while utilising your strengths?

CURRENT		FUTURE
I centralise governance efforts with a limited number of stakeholders with the understanding necessary to steer the organisation and its public image. The result is top-down decision-making, cascading communications and processes on an agreed scope of activities. This approach makes sense for some things, but not everything. More people must acquire the experience necessary to make the best decisions so information can flow faster.	→ **Develop** → ← **Utilise** ←	*The people with the best insight and decision-making ability are often closest to the customers, on the front line or even outside the typical organisational boundaries. Rather than controlling through process and hierarchy, we'll achieve better results by inspiring and empowering people at the edges to pursue the work as they see fit – strategically, structurally and tactically. We need to extend our governance efforts by developing a more engaged advisory network and community of advocates.*
Now you try:		

Finding our learning edge

It has become chic for leaders to say 'fail fast, fail often'. Failure has a spectrum, and some failures are detrimental to careers and organisations. We need to practise under less risky conditions to avoid failures where the stakes are high.

Snowboarders refer to 'catching an edge'. Catching an edge is when the edge of a snowboard accidentally digs into the snow, usually resulting in a fall or a near fall. These athletes learn how far they can lean over without falling. The only way to find that edge is to fall during practice. If you're engaging so cautiously that you never fall, you're not pushing yourself enough. How to fall safely is one of the first things to learn.

A new leadership refrain for exponential thinking should be: 'Fall, don't fail.' We must learn how to fall so that we can get back up and try again. See Table 1.10.

Table 1.10 Applied practice: think bigger

Consider a current data project. What support do you need to learn to 'fall'?		
What	**Who**	**How**
What do you see as possible? Where do you want to create bigger results?	**Who could help you develop and refine this vision? Who are your thought partners?**	**How might this vision become a reality? What mindsets and toolsets do you need?**
We need to determine and apply governance principles to AI customer service assistants, enabling them to give refunds to customers (just like our human customer service agents).	Engineering, customer service, fraud, legal, data science.	A clear understanding of customer history (buying and return habits) will provide helpful algorithmic training data. A clear understanding of what work tasks AI will do alongside and on behalf of humans as representatives of the company that aligns with our values, ethics and compliance guidelines.
Now you try:		

Key intervention and interference points

As mentioned earlier, humans constructively intervene and unwittingly interfere at different points in the data stack, depending on their level of awareness.

Data management can be hard to explain to people outside a data team. Even within the data team, not all functions are clear on how they support one another. Data roles cover complex data capabilities such as metadata management, data quality, architecture, cataloguing, privacy, data science and integration. Explaining underlying core concepts quickly and definitively can be a challenge.

Fellow data strategist Willem Koenders wrote an excellent article that included a figure which is re-presented in Figure 1.6 with his permission. He uses the analogy of real estate management to help illustrate how we might understand data management concepts.[28] Both require effectively organising, maintaining and utilising valuable assets. This helps not only to understand the underlying components, but also to imagine how they all operate together.

Let's walk through each of these comparisons:

- *Data asset*: At the heart of the analogy lies the data asset, which corresponds with the building or property in building management. The data asset can also be perceived as a data product or a dataset. Both data and building management revolve around managing assets that generate value when governed and nurtured adequately but lead to risks and losses when mismanaged.
- *Data (product) ownership*: A critical concept in data management is ownership – responsibilities may be delegated to others, but at the end of the day, one person or team should be the data owner. The same is true for a building, where this would be the property owner or landlord.
- *Data steward:* Data stewardship involves assigning responsibility for managing data assets to specific individuals or teams, for example, to ensure that data is of sufficient quality. In real estate management, data stewardship can be compared to the role of property managers responsible for the upkeep and maintenance of a property.
- *Data consumers/users*: Various individuals and business processes may consume the data, both internal and external to the organisation. This can be compared to the tenants that use the building for their respective purposes.
- *Data monetisation*: Data monetisation involves leveraging data assets to generate revenue, for example, by selling data to other organisations. In building management, this would be equivalent to finding ways to generate income from a property, such as renting space out to tenants or for an event, selling advertising space or selling the entire property.
- *Data contract*: A data contract is a formal agreement between a data producer and consumer, confirming what data will be exchanged and the corresponding formatting and quality requirements. This can be compared to a lease agreement, in which what is expected of the landlord and in what state the property will be available is described. It also outlines what the property can be used for (and specifically, what cannot be done to or with it) – the data contract can be used for similar purposes.

28 https://medium.com/zs-associates/the-best-way-to-explain-data-governance-to-beginners-c3b7a8feec15

Figure 1.6 Real estate as analogy to understand data roles (Source: Image used with permission: https://medium.com/zs-associates/the-best-way-to-explain-data-governance-to-beginners-c3b7a8feec15)

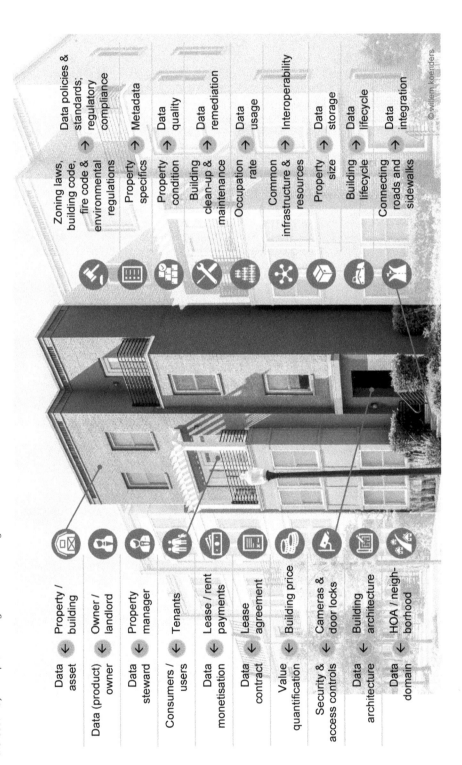

© willem koenders

Data asset ← → Property / building

Data (product) owner ← → Owner / landlord

Data steward ← → Property manager

Consumers / users ← → Tenants

Data monetisation ← → Lease / rent payments

Data contract ← → Lease agreement

Value quantification ← → Building price

Security & access controls ← → Cameras & door locks

Data architecture ← → Building architecture

Data domain ← → HOA / neighborhood

Zoning laws, building code, fire code & environmental regulations ← → Data policies & standards; regulatory compliance

Property specifics ← → Metadata

Property condition ← → Data quality

Building clean-up & maintenance ← → Data remediation

Occupation rate ← → Data usage

Common infrastructure & resources ← → Interoperability

Property size ← → Data storage

Building lifecycle ← → Data lifecycle

Connecting roads and sidewalks ← → Data integration

- *Value quantification*: For both real estate and data, estimating the value associated with the asset is worthwhile. Just as the value of a property depends on its location, size and condition, the value of data depends on its relevance, accuracy and accessibility.

- *Data security and access controls*: Data security protects data assets from unauthorised access, use or disclosure. In real estate management, data security can be compared to using locks, alarms and security systems to protect a property from theft or vandalism.

- *Data architecture*: This can be compared to the blueprint of a property, which defines the layout, design and construction of the building. Similarly, data architecture involves designing and structuring data storage and retrieval systems. Architecture standards can provide guidelines and best practices for building construction, and data architecture standards do the same for data assets.

- *Data domains*: Just as a city is divided into neighbourhoods, data can be divided into domains based on its subject matter. Any property belongs to a single neighbourhood, and together, all neighbourhoods include all properties – the same holds for data assets and domains. Each neighbourhood has unique characteristics, such as demographics and property values. Similarly, each data domain has unique attributes and requirements. An organisation such as a homeowners' association (equivalent to data domain owners or stewards) can be chartered to implement these requirements.

- *Data policies and standards, and regulatory compliance*: This can be compared to the different regulations that govern the use and development of properties, such as zoning laws, environmental regulations and building and fire codes. Similarly, data policies and standards define the rules for managing data in an organisation derived from applicable regulations such as data privacy and protection.

- *Metadata management*: Metadata is data about the data – it can describe the data asset in terms of the data attributes it contains, who owns it, who has access, who did access it and when, its location, how many records there are and the size of the total asset. Metadata can be compared to detailed information about a property and its features, for example, the total square and cubic footage, the owner, the number of rooms, its location and who has keys to the building.

- *Data quality*: Data quality refers to the fitness-for-purpose of data as measured along dimensions like accuracy, completeness and consistency. In building management, data quality can be compared to the condition and upkeep of a property, such as whether it has any defects or safety hazards.

- *Data remediation*: Data remediation refers to the process of identifying and correcting data quality issues. In real estate management, data remediation can be compared to the process of identifying and correcting property defects, such as a leaky roof or faulty foundations, to maintain the value and safety of a property.

- *Data usage*: This can be compared to the measurement of properties' usage, which helps determine their potential value. This measurement includes occupancy rates but perhaps even more detailed logs of who entered the building, when and for how long. Similarly, data usage measurement involves tracking and measuring how and by whom data is used in an organisation and to what extent data assets are adopted.

- *Interoperability:* This can be compared to a property's compatibility with other properties and (upstream or downstream) systems and its ability to share common infrastructure or resources. For example, a building is connected to the electrical grid, water network and sewage system, where each of these connections comes with precisely defined standards in terms of voltage, water pressure and pipeline sizes, and sewage standards. In a similar sense, data interoperability refers to the ability of the asset to exchange data and work together seamlessly with various other systems and applications, subject to accepted standards.

- *Data storage*: Data storage can be compared to a property's physical size and foundational structure. A property might have to be of a specific minimum size, for example, to accommodate industrial machines or to house families of a specific size. Similarly, data storage refers to the physical or virtual storage capacity in databases, data warehouses or data lakes.

- *Data lifecycle*: This can be compared to the lifecycle of a property, which involves various stages such as construction, maintenance, renovation and demolition. Similarly, data lifecycle management involves managing data through various stages: creation, storage, usage, archiving and disposal.

- *Data integration*: Roads and transportation systems connect different properties and neighbourhoods. A particular building may provide easy access to public transport and a nearby highway. Data integration involves connecting data from different domains and sources, including tasks such as data cleansing, mapping and transformation, to ensure that data from different systems can be used together. Without integration, you can't access or use the data, in the same way as you would not be able to enter or make use of a building.

The building management analogy provides a helpful way to understand the various aspects of data management and how they work together to support the organisation's overall data strategy. Now that we have covered human thinking skills and potential intervention points at a high level, it's time to understand the data supply chain.

The machine side

Now that we have explored the human side – *how* and *why* the data is used – we can start to follow data through the *when* and *where* of the inputs and outputs. Data is organised by layers commonly referred to as the data stack. Each data layer is a collection of components that solve discrete problems for a business function. When the layers are managed together as a stack, it establishes the foundation of an intentionally designed supply chain. Various data projects aid in developing and maintaining the data as a product and/or service. The ongoing strategic prioritisation of business requirements into data reports and insights is how data eventually becomes an asset.

As data moves through the stack, it goes through a predictable process – here, we'll refer to this as the data supply chain. There is a beginning (when it is acquired) and an end (when it is deleted and disposed of). During that journey, many human and machine activities change the makeup of the data as it moves along. There are three distinct layers in the data supply chain, as presented in Figure 1.7.

Once we move beyond the ownership mindset, the attention shifts to the details of the underlying data in systems. Working with data is complex and the data supply chain

Figure 1.7 High-level data supply chain (also called data stack)

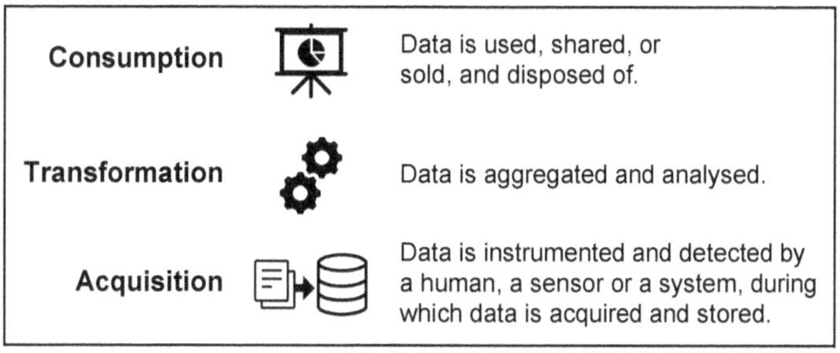

Consumption		Data is used, shared, or sold, and disposed of.
Transformation		Data is aggregated and analysed.
Acquisition		Data is instrumented and detected by a human, a sensor or a system, during which data is acquired and stored.

shows how interconnected every technical platform and data project is. Understanding each layer's purpose and activities, from hardware to data quality to analytical tools to business decision-making and so on – it becomes clear that *an organisation will perform as well as the lowest performing level of the supply chain.*

It doesn't matter if the reporting team has the most efficient processes. If the data sources are not managed well, if there is no single source of truth, those reports will become the subject of debate among executives (how are measures defined, calculated and so on?). When transitioning from a purely technical perspective to a data perspective, there must be minimum viability in all areas of the supply chain. This should prioritise projects from the lowest-performing area of the supply chain. This approach to viewing data and prioritising work represents a significant shift. Typically, more technical resources gravitate towards process improvement. They focus on doing things better, cheaper and more efficiently. Effectiveness might be the best way to engineer hardware, but it isn't necessarily the best way to manage data.

The following sections detail these three layers of the data stack. Each section contains a table with a description of the layer's function and examples of activities the data team typically considers at that layer. A corresponding figure summarises the data layer. There are also applied practice sections to test your knowledge.

Each layer of the data supply chain offers an opportunity for the data team to pause and reflect on how the data supply chain activities are impacted by incoming business strategy and governance requirements. Only then can they provide the context necessary for all stakeholders to consider the 'total cost' of supporting ongoing data and analytics for the organisation.

One of the biggest mistakes made by many companies is selecting a technology and then trying to retrofit everything around it. For example, a company was forced to use Vertica (an analytic database management software) because the CIO struck an enterprise deal with Hewlett-Packard. The client would have been better off with Snowflake or even open-source technology, but it didn't get that choice. It ultimately wasted several years trying to get its clickstream data to fit into Vertica before it was allowed to explore other options. Similarly, a CDO struck a deal with Snowflake in another company because they could not get the internal consensus needed to get all stakeholders into a central data warehouse. The five-year enterprise deal cost the organisation millions in unnecessary storage for datasets not actively used by the business.

When reviewing this section, keep those missteps in mind. Regard the data stack as distinct layers that serve particular functions. What does your organisation have or not have in each layer? What does it *really* need (and most importantly, *why* does it think it needs it)? Requirements and scope must align with business strategy to guarantee a return on investment.

Acquisition layer

Anything that gets analysed must be instrumented to collect data. Data collection is an intentional act that should occur during the planning phase of a business strategy, where (ideally) a data team member acts as a stakeholder in the business planning process.

Data comes into the supply chain as structured or unstructured. Structured data (that is, names, dates, addresses, credit card numbers, stock information, geolocation and so on) is highly organised and formatted to be easily searchable in relational databases. Unstructured data (that is, media and entertainment data, surveillance data, geo-spatial data, audio, weather data and so on) has no predefined format or organisation, making it much more challenging to collect, process and analyse.

Structured data (e.g. numerical/business data) is highly organised and formatted to be easily searchable in relational databases. Unstructured data (e.g. social media, multimedia, etc.) does not conform to any one particular standard and cannot be stored in relational databases since data strings have mixed data types which cannot fit into either a row or a column.

Both types of data influence decisions across the business in direction and indirectly. Structured problems (i.e. selecting office space) tend to be straightforward. They can be identified and solved by repeating a routine and decision makers can follow a definite procedure for handling them to be efficient. Unstructured problems present no clear definition, solution or process. They require judgement, evaluation and insights towards a unique solution.

Table 1.11 displays the data supply chain components featuring a critical instrumentation component and related actions.

Why does this matter?   When business requirements initially trickle in, most data teams absorb the initial cost. Lack of transparency not only masks the time and cost it takes to perform the task but also sows the seeds of misunderstanding that all data is free and easy to get.

Analysing unstructured data requires *significant* time and effort. Since unstructured data doesn't readily lend itself to traditional analytical techniques, humans (analysts or data scientists) must employ advanced methods such as natural language processing, sentiment analysis or machine learning algorithms. It has no predefined format or organisation, making it much more challenging to collect, process and analyse. Additionally, unstructured data requires a lot of storage space. It is expensive to store videos, images, audio and so on. Due to unclear structure, operations such as updating, deleting and searching can be complex. Storage costs are also higher as compared to structured data.

Table 1.11 Acquisition layer, instrumentation

Data stack component	Key activities	Value driven by the business	Organisational value driven by data teams
Instrumentation **Data must be instrumented to be collected from systems, sensors and humans. It focuses on where and how to get data.**	• Regular requirements-gathering sessions are conducted to understand long-term infrastructure needs. A shared strategy emerges from these sessions. • Instrumented data theoretically 'waits' for an event defined in the trigger to occur. When this event occurs, the customer moves through the activity from one campaign element to the next.[29] Each event is stored in a database. Data is stored and then deleted per the service-level agreement.[30] • Create log file standards. • Create a data catalogue. • Determine build or buy data catalogue plan.	• Once business requirements are determined, the business partners with the data team to consider the *context* of the data, *why* the data is collected, and if it *should be* collected. • When business stakeholders agree on what data gets prioritised and consumed from designated tools that manage a single source of truth, they can focus on the stewardship of driving business measures and accountability within their domain.	• The data team seeks to understand the business's key questions and help translate them into initial technical system and reporting requirements. • Increases ability to collaborate with other stakeholders. • Provides the ability to enable data access privileges. • Increases opportunity for partnership and platform, policy and process centralisation. • Reduces confusion, increases accountability, drives efficiency and productivity and reduces costs.

---

29 https://help.hcltechsw.com/commerce/9.0.0/management-center_customization/concepts/csbcustproflo.html

30 *All* data requested by the business must have an SLA in place and there must be an understanding of ongoing storage costs for the data. This is often overlooked, leaving IT to bear the costs alone.

Therefore, the full impact must be considered when helping the business translate problems into technical requirements for analytics and reports. Both the business and the data team must be able to (1) link the questions they are asking of the data directly to decisions that will impact the bottom line and (2) confirm the total cost of data cleansing, wrangling, storage (including archival, disposal and so on), so the data team might better prioritise and appropriately scope incoming requests.

Figure 1.8 illustrates the many kinds of data sources the business might want to monitor. This figure is not exhaustive but helps paint a picture of how the data team looks at and manages information coming into the supply chain.

The data team starts to plan activities such as data cataloguing, standardising log files and determining the platform and tools needed to accommodate the volume of data to manage. Both teams should concern themselves with the ethics of the data in which they are interested. While all data *can be* collected, teams should consider privacy and security implications, standards and data collection policies to confirm what *should be* collected. Use Table 1.12 to help you identify different kinds of data and explore their sources.

Applied practice: data acquisition

Acquiring data requires consideration of several attributes so that enough metadata (data about the data) is captured for it to be helpful in the future, such as:

- *Data source*: Where did the data come from?
- *Data structure*: What is the data's specialised format for organising, processing, retrieving and storing data?
- *Data samples*: What are the statistical analysis techniques used to select, process and analyse a representative subset of a population? For example, if mobile traffic is the source, was it on all mobile platforms or just one, was it just mobile traffic in Europe?
- *Data filtering*: What should be excluded, rearranged or apportioned (and to what criteria) at the acquisition point? For example, motion detection for security cameras or determining the total number of clicks per quarter.
- *Intended use*: What is the original intended use for the data (provides necessary context for analysis)?
- *Rights, permission and restrictions*: When gathering or using the data, what criterion should be considered for access, correction or portability of user data containing personally identifiable information (PII)? Data teams must be familiar with industry-standard privacy, security and trust policies or laws (for example, the EU GDPR or UK Data Protection Act).

Figure 1.8 Acquisition layer, big data sources

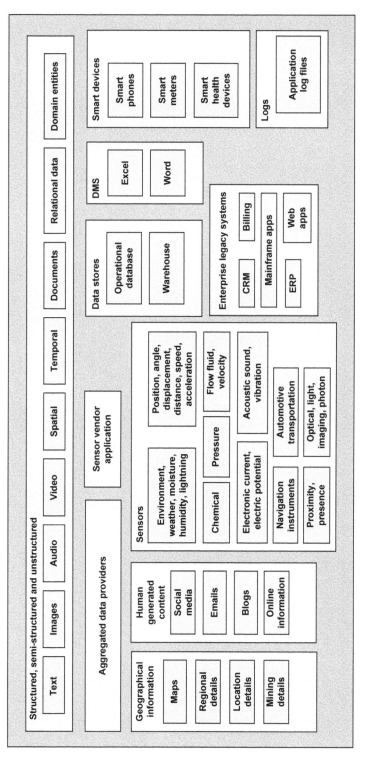

Table 1.12 Applied practice: data sources

Increase your awareness of different kinds of data and where they come from. Think of a type of data that surprises you or sparks your interest.			
Take a moment to search online for information about the types you selected. Make observations below.			
Data type (social media, financial, personal identity, health, business, generic and so on)	**Where could this data come from?**	**Who could be collecting this data?**	**What are the risks or ethical concerns in capturing this data?**

Transformation layer

Any data that gets analysed must be managed and stored. Data storage must be an intentional act, not an afterthought. Storage needs should be estimated during the planning phase of a business strategy, where (ideally) a data team member acts as a stakeholder in the business planning process. Are you sensing a theme of intentional collaboration developing? I hope so!

Table 1.13 displays the data supply chain components and key related actions.

Table 1.13 Transform layer, data aggregation, storage and analysis

Data stack component	Key activities	Value driven by the business	Organisational value driven by data teams
Aggregation **Keeps data in an accessible location**	• Establish taxonomies that define how the data fits together • Combine disparate datasets to create new value with big and little data; the sum is greater than the parts for analytics. • Determine core metadata fields;[31] determine build or buy metadata platform plan. • Determine data quality activities: checking for accuracy, ensuring formats are uniform between datasets and creating unique identifiers (to join data from different datasets). • Create alerts in data catalogues, updating stakeholders on what internal or external datasets are available. • Summarise the raw data to be displayed in reports based on business requirements. • Some transformations (cleansing, prepping and calculation activities) are automated in various operational processes and systems. • Develop service-level agreements for reporting efforts and data contracts from data provisioners clarifying the what, when and how of data delivery. • Both teams develop an idea of the cost investment needed to establish and sustain ongoing data reporting efforts. As the demand for data increases, so will the costs.	• The business owns the data. • When business stakeholders agree to govern a single source of truth, they can focus on driving business measures and accountability within their domain through effective data stewardship/ownership. Resources must be allocated accordingly. • The business is responsible for approving data policies and standards drafted by the data teams. • The business provides requirements for aggregation based on how they intend to use the data. For example, raw data can be aggregated over a given period to provide average, minimum, maximum, sum and count statistics. The data is then aggregated and written to a view or report.	• Data teams 'take care of' the data. • Data teams partner with the business to define the end-to-end processes and policies for data handling throughout the data supply chain. • Develops a 'single source of truth' for what and how data is used; populates and maintains a single data dictionary for the business. • Helps to understand, sort and validate datasets to increase their usefulness. Increases opportunity for partnership and centralisation, reduces confusion, increases accountability, drives efficiency and reduces costs. • The data team creates the environment to house the prioritised data as an asset through its systems-based practices. Prioritised data is the most meaningful data, driving decision-making. It contains insights that inform how business value is derived.

(Continued)

---

31 Metadata is data about another piece of data (its name, calculation, data source, etc.).

Table 1.13 (Continued)

Data stack component	Key activities	Value driven by the business	Organisational value driven by data teams
Storage **Saving data to a trusted location.** **The location is secure but still accessible for further manipulation and transformation.** **Storage can range from some version of the cloud to a specific server, a flash drive or local memory on a sensor. It needs to connect to other systems regardless of where data is stored.** *Note:* **The concept of storage is reconsidered after data is used. See 'Archival' and 'Disposal' in Table 1.20 in the Consumption Layer section.**	• Review of business requirements from business to determine what gets stored and for how long. • Review of business requirements to define parameters in the user interface for reporting tools. • Implementation or management of a central repository. • Monitor storage for performance. • Resource provisioning and configuration of unstructured and structured data, and evaluate how needs might change over time.	• Business teams can set prioritisation and take ownership of their respective domains. • Responsible for access control in partnership with the data teams.	• Ensuring data is protected and encrypted as it is moved to a trusted location. ▪ Reliable data preservation strategy ▪ Data continuity and accessibility ▪ Quicker and easier data recovery ▪ Provides security for protected files

(Continued)

Table 1.13 (Continued)

Data stack component	Key activities	Value driven by the business	Organisational value driven by data teams
Analysis **Asking questions about the past, present and future using data.** **After data is collected, stored and aggregated, it can be analysed. Data is examined to extract information and insights. Our analysis is often appended to original (or raw) data, or we might use it and discard extraneous data.**	• Summarise high-impact questions that drive the business. • Analyse the aggregated data to gain insights about resources or resource groups by: ▪ Stating and refining the questions. ▪ Exploring the data: past, present and future. ▪ Building formal statistical models. ▪ Interpreting the results. ▪ Communicating the results. ▪ Influencing decision-making, policies, practices and so on. • Develop a data quality framework to establish baseline and determine improvement over time. Determine quality thresholds and set alerts. • Determine standardised data quality dimensions.	Provides the data quality rules and/or business issues. Informs brainstorming and ideation sessions to address current problems and opportunities. Validates or disproves assumptions. Drives the prioritisation of business needs and generates more questions. Informs and confirms target customer personas and use cases. Drives testing and prototyping activities. Decision-making influences bottom-line measures such as (but not limited to) money made or saved, customer satisfaction and loyalty and organisational learning.	Consults and helps the business come up with standard metrics/rules. Ensures compliance with relevant regulations and policies (for example, partnership with a privacy team). Code and scripts are developed to meet use case needs through product features or services. Tests uncover defects which are analysed and fixed. Decision-making influences bottom-line measures such as (but not limited to) cycle time reductions, quality gained or lost, reduced risk and business value enabled.

57

Why does this matter? As data grows in volume, it must also grow in value to the business. The average company expects its data volumes to grow to 247.1 TB, an increase of nearly 52 per cent, in the next 12 to 18 months, with the plan to spend $7.93 million on data-driven initiatives in the next 12 months.[32] Yet, despite the increase in spending, most organisations expect their budget allocation for data-driven initiatives to remain the same in the next 12–18 months. With business requirements for reporting and analytics increasing and spending remaining (mostly) steady, data teams must help the business prioritise what gets into the supply chain.

Reporting frequency, for example, should be thoughtful. Each time a report refreshes its data, there is a cost. The business might request a report as a high priority because a vice president had a question. The requirement demands the report be refreshed throughout the day. It is forgotten about. A report set to be refreshed every 3 minutes over 12 months will drive up storage costs unnecessarily. The business stakeholders must justify their analytics requests by demonstrating a link between the questions they want to answer and the impact of the analytics while understanding the cost implications of their requirements.

Figure 1.9 illustrates the platforms and tools which the data team must use to fulfil the simplest requests. This figure is not exhaustive but helps paint a picture of how the data team manages and prepares data after it has been collected.

Figure 1.9 Transformation layer (aggregation, storage, analytics)

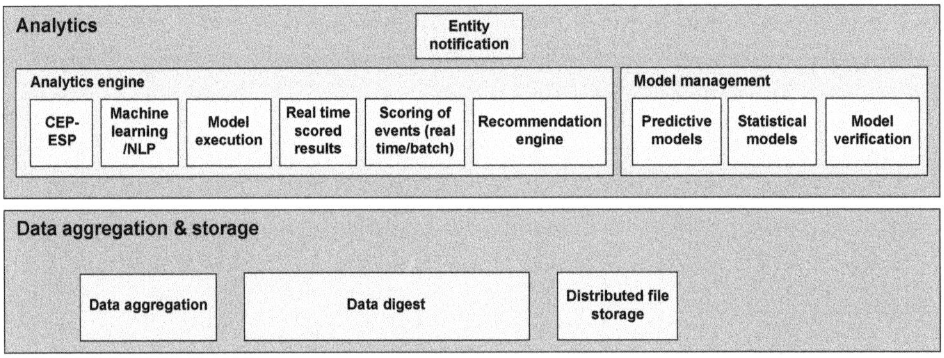

The data team starts to plan activities such as developing a comprehensive data dictionary, managing metadata and determining data access and permissions, and continues to determine more platform and tool needs. Both teams should concern themselves with the ethics of data storage and identify privacy and security implications, standards and data storage policies.

---

32 https://cdn2.hubspot.net/hubfs/1624046/IDGE_Data_Analysis_2016_final.pdf

Data aggregation   Aggregation is the combination of datasets to create a single dataset that's greater than the sum of its parts (often referred to as Big Data). Big Data combines as many data points as possible to help the company do better, faster and/or cheaper things. It is also where things get much more complex. See Table 1.14.

Table 1.14 Applied practice: is it aggregated and analysed?

Learn how data moves through the data supply chain.
Are you sure the data is accurate?
Is it standardised or normalised in some way that enables comparison to other datasets?
Have you created data catalogues that let you know what datasets are available inside your organisation or from partners? Are there taxonomies that define how that data fits together (such as finance, product, calendar taxonomies and so on)?
Does your organisation have standards for user privacy? Security? Trust?
What steps does your organisation take to anonymise data (such as a processing technique that removes or modifies personally identifiable information)?

Applied practice: data storage   Data at rest can be secured or 'locked up' until we need it – like using a filing cabinet. This static data stays the same until we choose to modify it. It is hard to synchronise static data across multiple systems, so people tend to keep copies of their favourite reports, adjusting them to their unique needs and producing multiple 'sources of truth'. Use Table 1.15 to think through real-world examples of how multiple data sources occur.

Data in motion is dynamic or 'live' – like water in pipes. The data must carry its security as it travels through and between systems. The 'pipes' must be secured in multiple ways – like locking the file cabinet in the prior example, and authenticating users accessing it. It is like signing a document and sealing it in an envelope before sending it through the mail.

Cloud storage is one way to simplify the problem of needing synced data to be accessible to numerous users in different teams or locations. Centralising data in the cloud helps to ensure everyone has access to the most up-to-date content without distributing every new version to every user. The term 'cloud' refers to the use of hosted services such as data storage, servers, databases, networking and software over the internet. While cloud storage and computing seem to remove geographic barriers to sharing and collaborating with data, local regulations about privacy and data use still may apply based on the physical location of servers and networks or the location of the discloser of the original data.

Data architectures must be built to synchronise and/or stream data across systems. However, even some tech-centric organisations didn't build infrastructures with the enormity of today's data in mind. Speed and robustness are hard to have ready in advance.

Table 1.15 Applied practice: how do multiple data sources occur?

Think about the metrics and data sources you use.			
Do you consume data from your reporting, other people's reports or directly from the data source?			
Measure	**Data source**	**Live or static?**	**Risks**
Renewals	*Finance*	*Static*	*Each sales department defines territory differently, so we take component data from finance and calculate it ourselves. There are five business units and five definitions of renewals; each has its own 'renewals report'. This prohibits the leadership team from determining whether to invest more in current or new programmes.*

Analysis   By this stage, data has undergone multiple forms of transformation on its journey through the supply chain. It has been gathered, catalogued, assigned, cleaned, joined, anonymised, regathered and reorganised – many human hands have touched it. To restate it another way, humans have interacted with data using existing mental models and biases (some constructive, some less so).

The analysis process turns data into information that can influence decisions and actions. Most data are essentially useless until analysis is applied. Consider the difference between the raw data generated by your banking transactions versus fraud alerts from your bank analysing transactions effectively.

Data analysis often falls into the categories of describing, diagnosing, recommending, deciding, predicting and/or discovering. Table 1.16 describes several analytics tasks, with examples.

Table 1.16 Analytics tasks with examples

Analytic tasks	Example
Describe what happened in the past or what is happening in the present	Determine last quarter's profitability or an electric vehicle's performance
Recommend options to humans or machine systems based on a variety of inputs	Suggesting movies or news stories to users based on what their peers have watched or read or proposing changes to text based on the tone a reader might expect
Diagnose the underlying cause(s) of a result	Identify which part is broken in a car's motor or find the source of new sales activity
Predict trends or potential outcomes with some degree of certainty based on past and present data	Forecast next quarter's sales or tomorrow's weather
Decide the next action without ambiguity – and without additional human input	Accept or reject a user's request for credit
Discover new opportunities or combinations within a particular domain	Find unidentified features in photographs, trending discussions on social media or commonalities between successful investment strategies

The analysis layer presents many ethical challenges – and opportunities.

As we have discussed, humans and machines carry out each step in the data supply chain. As we look at and are guided by the analyses from these numbers, it is important to remember how many hands have touched it along the way. The developers writing the code for them can unconsciously incorporate their understanding of what the business is trying to achieve. The people cataloguing and assigning metadata to them can be thinking with old mental models. For example, rationalising financial and calendar year hierarchies (where not both using Gregorian calendar) so users have flexibility in terms of how they view measures can be incredibly hard. Likewise, how we analyse and interpret meaning from these measures can add yet more bias into the mix, and those understandings, mental models and biases can be hard to remove, especially once a complex set of systems is in place. Metrics are a mathematical abstraction of what we value. Therefore, managing bias in the process is imperative when developing predictive analytics or machine learning algorithms.

Algorithms are like mathematical formulas. They are a step-by-step method for solving a problem, expressed as a series of decisions – like a flow chart or decision tree. Some advanced algorithms can 'learn' and update their decision trees using models that adapt over time. This is usually referred to as *machine learning*. Predictive analytics uses algorithms to predict the future based on the past. Machine learning is the more practical implementation of what is often called artificial intelligence. Natural language processing and sentiment analysis uses machines to understand how we communicate, that is, human language and meaning.

To avoid unintended consequences from data use, integrate active customer feedback loops. Use feedback loops whenever using algorithms that affect human beings to make sure the output of those algorithms is valuable and safe in the real world. See Table 1.17.

Table 1.17 Applied practice: learning to ask questions

Consider an example of big or little data related to your work. Imagine what information or insights you could get by analysing that data.
Example: *If I analysed the accuracy of data input and speed catalogued by time of day, I could look for patterns in how many errors were made in the morning versus the afternoon or error rates from notes taken in the field versus a dictation tool.*
If I analysed _____,
I could _____.

Why ALL data output (including AI) should always be regarded as the first pancake
Several threads on Reddit explore why people always burn their first pancake. While I won't delve into that here, pancakes make an excellent metaphor for AI output.

In determining the cover for this book, I was presented with the cockpit image shown in Figure 1.10. There were editorial and creative constraints regarding what kind of photograph I could use. When soliciting feedback on the image, some felt it was too dark and masculine and that the metaphor should have been a racing car, not a plane.

A former student of mine generated the image in Figure 1.10 using Microsoft Bing's Image Creator. He determined a simple prompt shown in that figure. Without much guidance, the image meets all the requirements of an engaging book cover and more. It is lighter, merges photography with abstract concepts and is generally more evocative. The figure is nonspecific on gender and ethnicity, rendering the image more inclusive. There is a clear mechanic/pilot execution emphasising knowledge *beyond flight*.

It is tempting to use this image; at least, I was tempted. Look what can be generated in less than two minutes with relatively little effort, compared to several conversations trying to capture the essence of over 350 pages.

Figure 1.10 Fun example of outsourcing a task to AI (Source: Created with Dall-E 2)

	Human	Human + Machine
Direction /Prompt	The title and initial Preface were submitted as the context for an initial image. There were back-and-forths where I suggested more abstract images (building blocks, scale icons and so on) and photos of drivers' dashboards and maps. Given the deadline and editorial constraints, the cover was determined.	'A photo image with mechanic as the analogy for data prep and visualisation and with pilot representing data environment'
Image output		

The image on the right looks professionally executed. Most responses from AI seem professional and read authoritatively. It reflects a reasonable answer to the prompt that was provided. However, we should consider it suspect. Why? Because we can't answer these (seemingly) simple questions:

- Where did it originate?
- Is it taking someone else's intellectual property?
- How should they be cited?
- Was the glow the figure is holding graphically added in or part of the photograph of the human figure?
- Did someone take that photograph, or is it mislabelled stock photography?

I'm no marketer, but I can execute a prompt. While I don't necessarily identify with a dark photograph featuring two white men, I do identify with the second image. The illustration overlays, the brightness and the emotional mood are more engaging to me – beyond what I could initially communicate to the editor for the initial creative brief.

This is a fascinating example of how we all need to uplevel our skills in the future. As someone leading a book project, I needed a vision for visually communicating my ideas, but I lacked the words. When the editor asked me for initial thoughts, I provided phrases and images to start the conversation. The graphics person took that as a starting point, along with the context from the book proposal, and started searching through licensed photography. We might have used AI to start the conversation and reach a different outcome.

Regardless of which image is better, there are several takeaways here:

- More tasks can be outsourced to AI than people realise.
- AI can do more synthesis, cheaper and faster than humans.
- Giving good direction (to humans and machines) is critical to success (however that is defined).
- All organisations must rethink how they will integrate, credit and utilise AI output in their future work.
- Everyone, and I mean *everyone*, will need to uplevel their data literacy skills.

Table 1.18 illustrates the importance and impact of learning to ask good questions. While everyone wants to know what is happening with revenue, how the question is asked directs the best solution. Every solution has time and cost considerations that most data teams (and sometimes IT teams) do not track and, therefore, absorb.

Table 1.18 Applied practice: learning to ask (good) questions

Consider an example of big or little data related to your work. Imagine what information or insights you could get by analysing that data.
Example: *If I knew why revenue was down, I could adjust the levers of my business (such as pricing basis, inventory allocation, product configuration and so on) with greater confidence.*
If I analysed _____,
I could _____.

Learning to ask *good* questions is where AI will shine a light on data services and data skills in the future. A typical analyst can't explore all the levers of a business in depth. Neither can AI, really. Asking questions (now called prompt engineering) is a critical skill everyone must master. It is about learning to steer a resource (human or machine) in a particular direction.

AI can synthesise non-conforming data (not just the data we intentionally code, but news articles, competitive data and other data fragments). To date, these kinds of data have not been considered business intelligence. However, Generative AI sees much further than human analysts can comprehend. That is the real breakthrough of future technology – that it can go beyond human capability and truly assist us.

AI can be dead right or dead wrong – and even failure can enhance our understanding, because it shows us where to do more investigation. We can now achieve a more multidimensional output than ever, but only if we understand how to ask good questions, how the data supply chain works and if data projects align with business strategy as a sponsored discipline.

Remember the idea that being a data consumer means understanding how data works? Asking questions is a mechanic's function, yet senior leaders must become more aware that their questions require specialised talent and incur costs.

Table 1.18 indicates that revenue might be dropping for several reasons. Asking a question that will return an answer for direct causation will point towards one solution. Asking a question that points towards statistical analysis will point towards another. Both analytics efforts require data science skills (expensive), big data storage (expensive) and time spent on reflection (potentially costly) and implementation (expensive).

AI learns from the nature and quality of our questions. It is trained off our skill set and biases for certain approaches, strategies and outcomes. What data language do we speak? How will AI adapt and amplify those blind spots?

Consumption layer

Analytics and insights are used, shared between various teams and eventually removed from the system. *What* data gets analysed is an intentional act – at least, it should be. Analytics requirements should be prioritised and agreed to by a consensus of business stakeholders during the planning phase of a business strategy, where (ideally) a member of the data team acts as a stakeholder in the business planning process. Unfortunately, *this* is when the data team is typically informed of what needs to be done.

Table 1.19 displays the data supply chain components and key related actions.

After data is used, storage is reconsidered once again. Table 1.20 lists last-mile storage considerations informed by governance and strategy.

Why does this matter? When data teams are merely *informed* rather than intentionally integrated into the planning process, they cannot effectively estimate the platforms and tools needed to support the business. As mentioned, data teams tend to absorb work until they burn out. When this happens, business units spin up competing data stack projects, ultimately increasing cost. While there might be a momentary boost to productivity, efficiency and data quality with multiple independent efforts by the business, this eventually decreases significantly over time.

Another issue with multiple immature data efforts is that it not only doesn't use resources efficiently, but also exponentially increases data storage costs because there is rarely a consistent approach to data disposal.

Table 1.19 Consumption layer, usage, sharing and disposal

Data stack component Definition	Key activities	Value driven by the business	Organisational value driven by data teams
Use Insights gained from data analysis are applied and used to make better decisions. These decisions affect change or otherwise help deliver a service or product. Data might be changed or appended during this stage.	• Applications of data inside and outside organisations • Datasets: providing collections of raw or lightly processed data points • Insights: deriving and providing new information from data • Algorithms: creating and training steps for machines to use in analysing other data • Optimisation: improving an organisation's operation or a user's experience • Products and offerings: improving an existing offering, such as a product or service, or creating a new offering • Investments: making decisions about where to apply time, money and focus • Personalisation: customising a user's experience with their data, third-party data and/or recommendations • Relationships: creating or improving human relationships through data	• Creating new business models • Evolving informed storytelling of the brand and business through insights • Develop an informed strategy for centralisation or decentralisation using data-based decisions • Determine strategic alliances, partnerships and acquisitions using deep insights	Develops APIs to increase customer engagement without building a lot of new infrastructure (for example, authentication APIs remember our passwords; they enable integrations of 'smart' technology or connected homes).

(Continued)

Table 1.19 (Continued)

Data stack component Definition	Key activities	Value driven by the business	Organisational value driven by data teams
	• Application program interface (API): a standardised way to pass data and commands between various systems to help them work together stably and securely • Selling or sharing datasets • Providing insights • Creating algorithms • Optimisation of operations • Creation of data-powered products • Making investments based on data • Personalising with little data • Creating and strengthening relationships		

(Continued)

Table 1.19 (Continued)

Data stack component Definition	Key activities	Value driven by the business	Organisational value driven by data teams
Share During the share or sell stage, we send data – or insights gleaned from data – back to its source or third parties. There are significant ethical considerations to consider at this stage.	Considerations for sharing and selling data. • Sharing data with the discloser: Instead of starting with data use cases for your business needs, ask what data could help your customers achieve their goals. • Sharing to improve quality, a customer service call centre might share recordings to receive coaching for its staff. • Sharing to deliver a service, as a personal budgeting app, might share account numbers with a bank to import transactions.	Considers how to share data back to its source – the users or organisations who furnished it – to create value and relationship by asking the questions: • How can data help users do something that already matters to them (such as predict travel time, develop smart shopping lists and so on). • Should we share or sell any data to other parties? • Should we share analyses back to the source of the data? • Should we ethically share or sell data? • Should we provide access to datasets – or the insights from those datasets, which are not the same thing – to other people, organisations or systems?	Considers the business requirement to share data back to its source – by asking the questions: • How should we share or sell data to other parties? • How should we share analyses back to the source of the data? • How should we ethically share or sell data? • How should we provide access to datasets – or the insights from those datasets, which are not the same thing – to other people, organisations or systems?

Table 1.20 Last-mile storage decisions

Data stack component Definition	Key activities	Value driven by the business	Organisational value driven by data teams
Archival The process of moving data that is no longer actively used to a separate storage device for long-term retention. Archive data consists of older data that is still important to the organisation and may be needed for future reference, as well as data that must be retained for regulatory compliance.	Establish a data retention policy Categorise data into types and then prioritise, carefully considering specific data required for ongoing operations and which data can be moved to the archive. Determine if data is stored in separate repositories or one centralised archive. Aligned with data retention policy or policies.	Business teams provide inputs on data archival requirements for specific datasets (for example, human resources needs to keep employee data for seven years).	Ensure policies are followed, reviewed and regularly audited.

(Continued)

Table 1.20 (Continued)

Data stack component Definition	Key activities	Value driven by the business	Organisational value driven by data teams
Disposal Data disposal is an essential consideration at the end of the data supply chain. When data is acquired, it must have a plan to be (eventually) discarded, recycled, reused or donated. Steps must be taken to ensure that the information stored in the data stack is either removed or 'sanitised' or completely deleted or destroyed. It's never the most exciting and doesn't generate any immediate value, so it's often overlooked. However, regulatory requirements and common decency demand that we consider how data will be disposed of when it's no longer helpful.	Data revocation and deletion strategies There are three main options for data destruction: (1) overwriting, (2) degaussing and (3) physical destruction. Example activities include: • Delete/Reformat • Wipe • Overwriting data • Erasure • Degaussing • Physical destruction (drill/band/ crush/hammer) • Electronic shredding • Solid state shredding • Sharing to provide insight, as a research app might share citations with a search engine to find related content. Data can be sold in several forms, whether raw, processed or as insights from data.	Business drives requirements related to data usage: • Sharing data with other organisations complicates the responsibilities of data disposal. When entering data partnerships, capture data disposal processes and standards in mutual agreements. • Data lasts forever until it is deleted. While people and circumstances change, how should a business respect users' 'right to be forgotten' if they need to revoke their consent later? • Determine if the data should be revoked from partners or purchasers, backups and other storage locations, and properly deleted or disposed of?	Data teams typically drive a federated versus centralised approach to managing data. From there, they ask: • How should data be secured? • How should data be disposed of when the business is done using it? • How should data be revoked from partners or purchasers, backups and other storage locations, and properly deleted or disposed of?

Figure 1.11 illustrates the platforms and tools available to various user groups. Data consumption is typically where self-service reporting sits. This figure is not exhaustive but helps paint a picture of how the data team provisions, shares and removes data after it has served its purpose.

Figure 1.11 Consumption layer (usage, sharing and disposal)

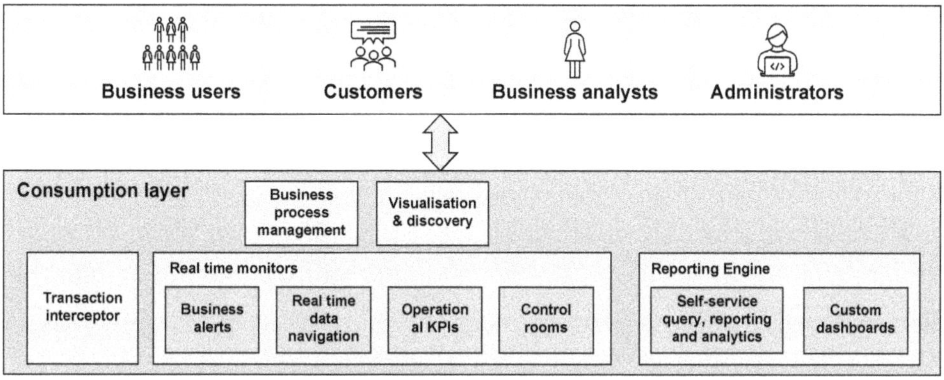

The data team starts to plan activities such as developing reports, scorecards and dashboards. The business team should lead monthly and quarterly reviews of its strategy, where the data team provides status updates on its ability to provide support through data instrumentation efforts, platforms and tools capabilities and so on. Both teams should include ethics discussions on standards and policies for ongoing data access and disposal.

A data team's ultimate insurance policy is its data cultural and literacy effort. To help drive awareness and cultivate the necessary mindset shifts needed to increase collaboration between business and data teams, everyone must have a common understanding of the data supply chain and their role in it. They must also understand the human and machine partnership required to get the best data out of their systems. Data literacy ensures that each user group understands the available data services, how to use them and how to apply data concepts in their daily work.

The value driven by business, IT and data teams is not entirely black and white. Most of the *Should We-versus-How Should We* conversations are collaborative discussions. In general, business is thinking about creating features and selling experiences (via products and services), and the data team is thinking about enabling and building those experiences in the most trusted way possible. However, the data team (regardless of where it sits) considers how to prioritise, curate and manage the data required to power the organisation's operating model.

Connecting human + machine: putting it all together

So, we have explored some big questions of the data supply chain: What does it take to design a data infrastructure system? What employees get access to what data? How are they trained, and how is data integrated into ongoing performance? What kind of norms do infrastructures enforce, and which employees are enabled to thrive? What happens when infrastructure breaks down or proves inadequate in the face of a crisis? What do infrastructures teach us about our decision-making? And what kind of world do they make possible?

Answers to these questions are informed by *human* activities such as developing strategies, determining ongoing operational activities and enforcing governance. Mental models of culture, rank, function, ability, leadership and functional and organisational maturity influence every design decision in the company – including decisions about data discovery, transformation and consumption. Example questions that drive data infrastructure design include but are not limited to:

- Do we centralise, federate or decentralise data management to enable ongoing business decision-making? Why? Why *now*?

- Do we recruit talent with depth in problem-solving skills to enable an agile workforce? Why? Why *now*?

- Do we automate decision-making with machine learning to reduce operational drag in the system? Why? Why *now*?

Data teams should add these questions to their list when considering the people, processes and values that shape the data infrastructures needed to meet ongoing requests from the business. How these systems simultaneously supply opportunities for and place constraints on data access, sharing, analysis and insights that enable effective business decision-making exposes the organisation's (and leaders') blind spots, biases and politics.

A common misconception of early data teams is that the work is done once an initial data team and data stack are set up. Experience tells us that reports have short life spans, because the questions improve. Better questions mean more labour and expertise are needed to mine, prepare and analyse the data.

Data must be *maintained*. Requirements from the business will only increase. Therefore, support to the data teams supporting these efforts must also increase. Everyday ongoing activities include:

- Data strategy and requirements gathering (in response to the business strategy sessions).

- Integration (protocols, connectors, adapters, native APIs, connectivity (offline/on-demand)).

- Systems management.

- QOS layer (frequency, availability, size/tech, security, filters, privacy, exception handling).

- Governance.

- Data literacy.

Figure 1.12 shows an example of a data stack, incorporating the key concepts of acquisition, transformation and consumption of their related activities. This kind of diagram can spark good conversations between the data team and their stakeholders about the need for log standards, data sources and ownership, data modelling and the ongoing sustainment processes that keep the stack relevant to stakeholders.

Figure 1.12 Sample data stack with key elements

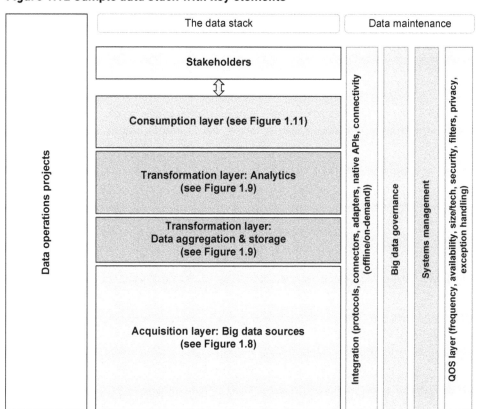

A word about bias

Analysis presents many ethical challenges – and opportunities. Ultimately, bias comes down to the data and the algorithms we use. Algorithms codify and amplify any biases in current processes or problems with historical data. Data teams must help their business stakeholders to think through the implications of their strategies. Generally speaking, technical teams tend to be less representative of business concerns than their stakeholders. However, they must ensure their business counterparts have considered every potential risk and consequence.

For example, the retail company Target exposed a teen girl's pregnancy by noting a change in her buying habits.[33] A data scientist translated the 'new parent' or 'pregnancy' business requirement as a binary variable with logistical regression. This approach is standard for many types of advertising targeting. The result presented pregnancy products to a 13-year-old girl. The technical implementation should have been handled differently. Decisions of methods, data collection processes and privacy protocols in service of these business requirements happen at the project level, not the executive level.

---

33 https://www.nytimes.com/2012/02/19/magazine/shopping-habits.html?_r=1&hp=&pagewanted=all

Another example of how data teams must help business stakeholders think through their requirements came from a 2023 report by the non-profit Mozilla Foundation.[34] It found privacy and security concerns among the top 25 automakers, saying they received failing marks for consumer privacy. The study found that automakers are collecting more data than necessary – not just how fast or where we drive, but the demographics of drivers and passengers. Vehicles have been collecting information on drivers for some time. These concerns should have been vetted much earlier. Even without regulations, the automakers should have anticipated that regulations would come. What is permissible today will likely not be permissible tomorrow. It is extremely easy for leaders – at every level – to abdicate their responsibility and ask, '*What does the data tell us to do?*' Data is so prevalent in our lives and work that we can sometimes assume that all of it is objective when it is simply a number or suggests a declarative insight. Remembering the human side of the supply chain (the how and why the data inputs are made) and the machine side of the supply chain (the when and where of the inputs and outputs) should amplify the many areas where data is decidedly *not* objective.

Bias is part of the process.

Awareness of how the organisation's values, leadership priorities and market pressures shape our interventions with data is critical to mindfully working with data.

Too often, algorithms are thought of *only* as mathematical formulas. As mentioned, they are mathematical expressions of human values (like growth). Business might define it one way, but the developers translating a business definition into code brings unconscious biases. Those biases can be hard to remove, especially once a complex set of systems is in place. Therefore, when developing predictive analytics or machine learning algorithms, a process must be ensured to manage bias.

Two very different practical examples illustrate how bias enters a system. The first is a typical example of the importance of stakeholder agreement on the common understanding of measure definitions. The second is the egregious impact of racial bias in automated processes. Both can result in lawsuits, embarrassment and unintended consequences.

Case One: Implications of incorrect metric definitions

Before online measures were generally determined and agreed to across most internet companies, Nielsen (an internet ratings body), Microsoft (a consumer software company) and Omniture (a web analytics company) tried to reach a common understanding of the definition of a pageview. The definition from the business was 'an instance of an internet user visiting a particular page on a website'. The engineer interpreted the definition, wrote the code and presented a sample report. However, business, legal and engineering still had three different

---

34 https://www.cbsnews.com/losangeles/news/new-study-finds-that-automakers-collect-too-much-personal-information-about-drivers/

totals because each had been tracking their version of pageview (and defining them differently). The engineer used an invisible variable in the code, such that when the page *started* to load, a 'tick was allocated to a pageview total'. Business and legal noted that the tick was allocated before the whole pageview was resolved, impacting ad banner performance (revenue) and legal disclaimers (third-party agreements). The engineer had no operational need to see the whole page through, so they had logically placed the invisible variable at the *top* of the page when the need for it was at the bottom.

Key takeaway: There are VERY large websites that still have their tags at the top of the page and probably shouldn't for this very reason. This example illustrates the significance of business context for a definition.

Case Two: Implications of gender bias in algorithms

Goldman Sachs, which hosts the Apple credit card infrastructure, developed an algorithm that screened applicants for creditworthiness. In certain situations, its algorithm somehow erroneously concluded that women were less creditworthy than men. In one notable case, a couple with completely joint finances received very different offers. The male member of the couple received a credit limit 20 times higher than his wife's despite her superior credit score.[35]

Key takeaway: It was a scandal, and with the two firms' interdependence, it was harder to identify who was responsible.

We are rethinking many aspects of data management as an industry that are no longer permissible. Just six months ago, something might have been okay for the simple reason that no one had yet pointed to it as a bad idea. We think we can use our judgement but, even if we can figure out a technical implementation that drives value, such as the example with Target, we must consider the impact on the data architecture if regulations were present.

The answers to business requirements are never black and white. Should the auto companies collect data on their drivers? If so, how much? Where should it be stored? For how long? These decisions are made against a continuously moving target. Should we move forward with facial recognition? Maybe. It's still better than nothing, but it's less accurate with darker skin tones. Data teams must think about future technology shifts in their data architecture planning. The number of things we *can* achieve with new technology will always increase, just as the things we *should* do tend to decrease over time.

Regardless of a metric's complexity, when the context is not transparent, it's challenging to do forensic analysis to determine the source(s) of the unintended outcome. Was the code wrong, or was the business definition not specific enough? Was it the accuracy of the data fed into the code or algorithms? Or was the code or algorithm itself wrong?

35 https://www.washingtonpost.com/business/2019/11/11/apple-card-algorithm-sparks-gender-bias-allegations-against-goldman-sachs/

In both of the cases described above, identify and/or mitigate the unintended consequences of data use. Recommendations coupled with questions such as 'Did we make the right recommendation?' can help teams and the writers of algorithms make the right course corrections.

A crucial part of increasing data fluency is finding ways to help humans increase awareness of their impact and more intentionally guide the direction of machine processes and algorithms in a more accurate way.

As stated earlier, this bears repeating: *it is still incumbent on the data consumer (generally a data team and their stakeholders) to know how data works. We can't let any data solution turn into a black box. Whenever we receive output from AI or ML, we must have the skills to challenge it. We must consider such output as 'the first pancake'. We must examine the output, learn how to challenge the data sources, and undertake the overall preparation to defend our logic with an answer that aligns with a more perfect pancake.*

Data management use and need today

Investing and curating data as a service is becoming an increasingly important part of how organisations build and keep their advantage, especially as the need for accurate, timely data increases and financial resources remain tight. Data infuses various organisational functions – most commonly in finance reporting, but not exclusively. Moreover, creating data as a service is the first step towards curating a corporate asset. Table 1.21 illustrates the degree to which organisations like yours are using data to drive fundamental decision-making.

Table 1.21 Data used by organisations to drive decision-making (Source: Softjourn, https://softjourn.com/insights/data-driven-decision-making)

Finance	60%
IT	41%
BI Competency Center	41%
Marketing	24%
Other business departments (human resources, etc.)	8%

Data-driven decision-making already impacts different types of businesses across every industry, including Fortune 500 companies. According to research from the IDC, global spending on big data reached $215 billion in 2021, an increase of 10 per cent from the previous year.[36]

Table 1.22 highlights the countries currently making the most significant investments in big data and analytics are banking, IT, discrete manufacturing, professional services, process manufacturing and telecommunications. The biggest markets for big data and analytics are the United States, Japan, China and the United Kingdom.

---

36 https://www.idc.com/getdoc.jsp?containerId=IDC_P33195

Table 1.22 Top countries based on market share (Source: IDC, Worldwide Big Data and Analytics Spending Guide, https://www.idc.com/getdoc.jsp?containerId=IDC_P33195)

USA	51%
Japan	28.8%
Germany	5.5%
United Kingdom	5.1%
People's Republic of China (PRC)	4.4%
Others	5.7%

Data leaders must navigate multiple transformations to deliver the capabilities of a successful, efficient and agile organisation in the digital-first economy. IT/technology, finance and operations are generally most involved with leading and scaling resiliency across the organisation. But functions like human resources and marketing make excellent strategic partners and can do much to support the effort. To achieve this mission, data teams need to fulfil four crucial roles:

- architecting, orchestrating, integrating and managing ecosystems of digital technologies and assets from disparate and evolving sources;

- building organisational resilience, agility, scalability and security;

- leading and governing the evolution of enterprise and ecosystem digital infrastructure, products, services and experiences;

- creating communities of practice that deliver sustainable streams of custom software and novel digital technology applications and solutions.

Data-driven organisations must increase their awareness by assessing their current culture's competency, maturity and ability to support a digital-first mission. The traditional approaches of strategy and planning and the annual production of a three-year plan that worked in a predictable context have become obsolete or misaligned with the uncertainty data teams face. There is a need to move towards more adaptive approaches to data projects and infrastructure planning and implementation, enabling quicker responses to customer feedback.

Not everything will change overnight. Not every process or technology needed for success will be new. There is only a 'here' and a 'there', with constantly adjusting deadlines. The future will always be a hybrid of what is complicated but known and what is complex and unpredictable. Success rests on our fluency, adaptability and ability to learn dynamically, *moving between these paradigms*.

Despite the wide use and supply of talent, data organisations in the research conducted with Forrester still reported a significant gap between the data skills in the current workplace and what is needed.[37] Data skills need to expand across all aspects of business. Once required only by specialists, data skills are now necessary for workers across the enterprise. Employees in Forrester's study across every department – including product,

---

37 https://www.tableau.com/sites/default/files/2021-06/Tableau_Data_Literacy_Report.pdf

IT, HR and operations – cite digital skills as the most essential in work and life. In which of these functions could *you* use better data management? See Table 1.23.

Table 1.23 Employee demand for digital skills by discipline (Source: Forrester, Nov 2021)

Discipline or field	Need
Customer insights	89%
Analytics	85%
Finance/accounting	80%
Human resources	86%
IT	90%
Marketing/advertising	87%
Operations	86%
Product	92%
Research	83%
Engineering	81%

Here are some examples of the possible range that data projects can have. These projects were immediately valuable for the organisation, and they created a variety of related or downstream benefits. They range from projects done in a day to projects lasting a year or more.

Examples of being data-informed

- *Data-driven decision-making*: Google created its People Analytics Department to help the company make HR decisions using data, including deciding if managers make a difference to their team's performance. The department used performance reviews and employee surveys to answer this question. Initially, it appeared as if managers were perceived as having a positive impact. However, a closer look at the data revealed that teams with better managers not only performed better, but were happier and worked at Google longer.[38]

- *Increase the relevancy of (product) recommendations*: Amazon bases its recommendations on what customers have bought in the past, the items in their virtual shopping cart, what items the customer has ranked or reviewed after purchase and what products the customer has viewed when visiting the site. Amazon also uses key engagement metrics such as click-through rates, open rates and opt-out rates to further decide which recommendations to push to which customers. By integrating recommendations into nearly every aspect of Amazon's purchasing process, from product browsing to checkout, the company has found that product recommendations drive sales and increase the bottom line.[39]

- *Boost customer loyalty programmes and optimise operations*: Southwest Airlines uses customer data to decide what new services will be most popular with

---

[38] https://www.smartdatacollective.com/analytics-google-great-example-data-driven-decision-making/

[39] https://fortune.com/2012/07/30/amazons-recommendation-secret/

customers, and most profitable. Southwest has found that by observing and analysing customers' online behaviours and actions, the airline can offer the best rates and customer experiences. As a result, Southwest has seen its customer and loyalty segments grow year after year.[40]

- *Redesign interface to improve customer experience*: Airbnb built its own A/B testing framework to run controlled experiments to inform decisions around product development. Data could not generate the idea for the redesign in the first place; nor could it (on its own) make clear why it initially seemed to perform poorly. Human guidance was needed to interpret the context of the experiment results. Data is there merely to help.[41]

- *Unlock customer patterns*: Disney uses data mining to examine individual visitors' prior behaviour and preferences, and forecasting algorithms to estimate the types of holiday packages they are most likely to enjoy.[42]

- *Create relevant content to increase customer retention*: Coca-Cola built a strong strategy around data analytics as the main factor in achieving customer retention. Data plays a vital role in the marketing and product development of the label, helping the Coca-Cola team create relevant content for various audiences and following consumers' input to adjust the team's approach. Data informs the brand's sourcing, distribution, sales, production and other decisions. According to Greg Chambers, Global Director of Digital Innovation, data analytics and AI 'have been woven into the fabric of Coca-Cola'.[43]

- *Improve customer satisfaction*: The industry status that Netflix has achieved is due to its insights showing what most commonly triggers people to subscribe, what makes them stay longer and where best to invest further effort in improving its services. Also, by using past search and watch data, Netflix can direct its purchasing of new content, improve suggestions to its subscribers and improve customer management and satisfaction.[44]

- *Address common customer issues fast*: Uber stores data for every trip, with or without customers. The company also looks at how transportation happens in different cities and countries and uses those insights to make predictions and improve services. Through data analytics, Uber addresses the most common issues for its customers and acts to remove them. The company also uses this method to adjust its supply and demand strategy and accelerate its influence on the market, topping competitors along the way.[45]

A data oath

'When you see something that is technically sweet, you go ahead and do it', Oppenheimer told a government panel that would later assess his fitness to remain privy to US secrets. 'And you argue about what to do about it only after you have had your technical success.

40 https://fortune.com/2014/06/19/big-data-airline-industry/

41 https://medium.com/airbnb-engineering/experiments-at-airbnb-e2db3abf39e7#.miqyczkzb

42 https://www.appventurez.com/blog/how-big-data-and-machine-learning-drove-disneys-magic-globally

43 https://d3.harvard.edu/platform-digit/submission/coca-cola-leverages-data-analytics-to-drive-innovation/

44 https://www.engati.com/blog/predictive-analytics

45 https://neilpatel.com/blog/how-uber-uses-data/

That is the way it was with the atomic bomb.' His security clearance was revoked shortly after his testimony, effectively ending his career in public service.[46]

That was in 1942.

In 2023, Geoffrey Hinton, the godfather of AI, officially joined a growing chorus of critics who say that companies are racing towards danger with their aggressive campaign to create products based on generative artificial intelligence. This technology powers popular chatbots like ChatGPT. Dr Hinton quit his job at Google, where he had worked for more than a decade and after having become one of the most respected voices in the field, so he can freely speak out about the risks of AI. A part of him, he says, now regrets his life's work.[47]

Today is our Oppenheimer moment.

We are only *beginning* to reckon with the certainty that automated decision-making systems handle some of our most important organisational, political and societal decisions. In many ways, the data science community is experiencing a slow-burn ethical crisis. How will AI impact healthcare, the military and education?

While a professional code of conduct should be an internal, thoughtful effort led by the people who are actively doing data science day to day, the people affected by the most invasive and automated decision-making systems of our time (employees, customers, patients and so on) also have valuable insight to offer.

A user experience lens to working with data is critical to developing the most holistic, human-centred and ethical solution. We can no longer consider ethical, long-term data management an afterthought of business strategies. In her book *Mismatches*, user experience expert Kat Holmes writes:

> Mismatches are barriers to interacting with the world around us. They are a by-product of how our world is designed. Mismatches are the building blocks of exclusion. They can feel like little moments of exasperation when a technology product doesn't work the way it should. Or they can feel like running into a locked door with a big sign that says, 'Keep out'. Both hurt.

Virginia Eubanks shares stories of how the digital tools we interact with worsen bias, subvert democracy and violate our fundamental human rights. In her book *Automating Inequality*,[48] she says that these experiences and the many security breaches and data privacy issues faced by leading organisations suggest that we need to think more broadly about who is included in and excluded from the 'community' of data science. These systems affect us all as citizens, employees and consumers – but they don't impact us equally. Without including the voices and experiences of those who face automation's most dire effects, data scientists may miss the opportunity to push the field towards social justice.

---

46 *In the Matter of J. Robert Oppenheimer, USAEC Transcript of Hearing Before Personnel Security Board* (1954).

47 https://www.nytimes.com/2023/05/01/technology/ai-google-chatbot-engineer-quits-hinton.html

48 Eubanks, V. (2018) *Automating Inequality: How High-Tech Tools Profile, Police, and Punish the Poor.* First edition. New York: St. Martin's Press. https://virginia-eubanks.com/automating-inequality/

Oath of digital non-harm, from *Automating Inequality*

I swear to fulfil, to the best of my ability, the following covenant:

I will respect all people for their integrity and wisdom, understanding that they are experts in their own lives, and gladly share all the benefits of my knowledge with them.

I will use my skills and resources to create bridges for human potential, not barriers. I will create tools that remove obstacles between resources and those needing them.

I will not use my technical knowledge to compound the disadvantage created by historic patterns of racism, classism, ableism, sexism, homophobia, xenophobia, transphobia, religious intolerance, and other forms of oppression.

I will design with history in mind. To ignore a four-century-long pattern of punishing the poor is to be complicit in the 'unintended' but terribly predictable consequences that arise when equity and good intentions are assumed as initial conditions.

I will integrate systems for the needs of people, not data. I will choose system integration as a mechanism to attain human needs, not to facilitate ubiquitous surveillance.

I will not collect data for data's sake, nor keep it just because I can.

When informed consent and design convenience come into conflict, informed consent will always prevail.

I will design no data-based system that overturns an established legal right of the poor.

I will remember that the technologies I design are not aimed at data points, probabilities, or patterns, but at human beings.

      _____   _____

                    Name   Date

Take a moment and reflect on what you have learned about the data supply chain. Recall the many opportunities to constructively or unconstructively intervene with data. Think about the impact you have or can have on your company's data integrity.

Take the oath. Think about how you want to lead data projects with more intention towards a better, more informed society that benefits everyone.

KEY POINTS

- Data is a rational investigation of the truths and facts/information, which must first go on a journey and experience transformation.

- Terminology is shifting, causing much static in the data industry. It is more important than ever to be precise when describing processes, requirements and policies and determining naming conventions.

- For data to create and generate value, everyone in the organisation needs to increase their skills in order to use data effectively to drive decisions and actions. This requires a general understanding of the human roles in data management and the supply chain.

- The human side of the data supply chain covers *how* and *why* data inputs are made. The machine side of the data supply covers the *when* and *where* of the inputs and outputs.

- To become the best partner to the machines we rely upon, we must learn to embrace a few thinking principles that will stretch us from our current comfort zones. Specifically, we must challenge mental models, think computationally, increase our ambidexterity and find our learning edge.

- Data roles cover complex data capabilities such as metadata management, data quality, architecture, cataloguing, privacy, data science and integration. The analogy of real estate management helps illustrate how we might understand data management concepts. Both require effectively organising, maintaining and utilising valuable assets. It not only helps to understand the underlying components but also to imagine how they all operate together.

- As data moves through the stack, it goes through a predictable process – the data supply chain – where data is acquired, transformed and consumed.

- Human activities such as developing strategies, determining ongoing operations activities and enforcing governance inform the data supply chain.

- Mental models of culture, rank, function, ability, leadership and functional and organisational maturity influence every design decision in the company – including decisions about data discovery, transformation and consumption.

- Data must be *maintained*. Requirements from the business will only increase. Therefore, support to the data teams supporting these efforts must also increase.

- Everyone working with data should consider an oath to do no harm as part of their organisational ethics.

2 DATA TRANSFORMATION 101

Those able to make the most progress fastest stand to capture the highest value from data-supported capabilities. Reducing the time it takes to respond to data requests can generate cost savings of 30–40 per cent; organisations that simplify their data architecture, minimise data fragmentation and decommission redundant systems can reduce their IT costs and investments by 20–30 per cent.[49]

When business stakeholders provide increasingly sophisticated data requirements to a data team without intentionally considering the underlying architecture required to support them, they add significant drag to operations. Data leaders at every level must take every opportunity to educate their stakeholders about *everyone's* responsibility to nurture the data supply chain that enables the organisation's competitive edge.

As a leader of data transformation efforts, you are deeply committed to your vision and making an impact. Guiding initiatives through a pandemic, global unrest and economic turmoil requires doing more with less. Yet, you also realise there is a limit to what you can do with current resources. The organisational culture must shift and learn to adapt quickly. But how?

The need to not just collaborate but *intentionally integrate data into business strategies* is essential to maintain the speed of business. This insight makes you a pragmatic optimist: you see the limited potential of what can be achieved on your own through ad hoc efforts, realise the exponential benefits of intentional interdependence with key stakeholders – and know that you need to be scrappy to achieve it!

One of the most important resources you need is for everyone to intimately understand the human and machine aspects of the data supply chain – and their responsibilities in that supply chain. This awareness helps ease the friction that accompanies collaboration. It also points towards an approach that will strengthen your organisation's structure and upskill everyone in the process. Most organisations can't pay for consulting expertise in every function. But even if you can't afford to bring in high-priced consultants for different projects or advising roles, you can still invest in developing yourself and your team by evangelising a common understanding of how data becomes information (and everyone's role in that process). You can also increase the rigour of how data projects are led. Not everyone needs functional certification or a graduate school degree to execute data projects well. But they do need a path forward. They need to know where they are in the data supply chain to determine the most strategic approach for executing data

---

49 'Designing a data transformation that delivers value right from the start', October 2018, McKinsey & Company, www.mckinsey.com. © 2023 McKinsey & Company.

work to contribute to a corporate asset, and they must strive for constant alignment of tactics and strategy.

To do these things well, we need to challenge old mental models as we move towards new ways of executing work. We need to understand how data flows and transforms through a system to create value – and build new business models around that new value. We need to evolve technical, interpersonal and leadership skills. Cultivating data as a corporate asset helps people and organisations to grow exponentially.

Further, developing a transparent approach towards leading data projects helps better manage the human interactions we take for granted that directly impact data quality. In many ways, a solid strategy for project management, where we learn to break problems down into component parts, is the first step towards learning to think and work like a machine. But we can't forget the human side – the awareness (or lack of it) of what needs to change, the desire (or lack of it) for change and the ability (or lack of it) to utilise what we develop. Tending to the human side of data projects impacts the quality of the data requirements we receive, the level of engagement for ongoing stewardship and the understanding of compliance efforts, policies and ethical concerns of governance initiatives. Tending to the human side of data projects makes us more mindful of our impact on the data we acquire, manage, store and sell – to name a few things that data influences!

Breaking down the problem of data transformation into smaller components will demystify and reduce some of the overwhelm that often accompanies these initiatives. When reviewing the smaller components, there might be a tendency to oversimplify or take for granted some of the effort which smaller projects might make. This is where a solid understanding of the data supply chain (and corresponding activities) and the rigour of the approach can help to keep you on the right path.

DRIVEN BY DATA AND INFORMED

Debate regarding the phrase 'data-driven' versus 'data-informed' arose in attempts to mitigate unintentional overreliance on data. When relegated to buzzwords and jargon, these phrases infer that being data-driven means the decision-maker is on autopilot, abdicating their subjective knowledge of leadership, the market and their own experience. Such a stance, everyone can agree, is a misuse of data (and the term).

The fundamental difference between being data-driven and being data-informed is that being data-driven lets the data guide the decision-making process. The data acts as the arbiter for decision-making. Taking a data-informed stand allows data to check intuition and subjectivity.

As illustrated in Table 2.1, both approaches are valid. There is a time to be data-informed and a time to be data-driven in our decision-making. For example, using past and current data, a data-driven approach can help us predict the future. Without data, we risk making incorrect assumptions and being swayed by biased opinions.

Table 2.1 Data-informed and data-driven decision-making: advantages and disadvantages

	Data-informed	Data-driven
	Leverages exponential thinking and invites us to use data to think differently and creatively	Leverages computational thinking and lays the path to machine learning, AI and automation
Advantages	Get 'big picture' perspective	Automation of decision processes
	Complement data with other decision inputs	Reduce or eliminate partisan influences
	The value of experience in decision-making	Reduce decision cycle time
	Creative solutions for complex situations	Capacity for frequent high-volume decisions
	See disruptors not visible in historical data	See disruptors not visible in historical data
Disadvantages	Easy to introduce biases	Missing the big picture
	Influence by the most powerful or most vocal	Potential for biases in algorithms
	Cherry-picking of data (confirmation bias)	Risks of poor data quality
	Complexity of too many inputs	Risks of unreliable data pipelines
	Difficulty explaining decision rationale	Analytic models may decay over time

Using the two data approaches together will increase the relevant insights your teams require to do their jobs. The key is to understand when to use each so that your expectations of the data match what you will get from it.

Both mindsets have a place. The key is knowing which ones are required and when, and moving fluently between them.

MAKE AN IMPACT: RAISE ALL BOATS

While a fully funded budget that supports data as a service is an integral part of the financial picture for an organisation working with data transformation, few organisations are fully staffed or funded. Three-quarters of executives confirm their organisation now has some form of data strategy (however rudimentary), but a paltry 16 per cent say they have the skills and capabilities necessary to deliver it.[50] Even though the average staffing budget is growing yearly, finding the skills and capabilities to execute data projects is becoming harder and harder.

Many organisations start with ad hoc efforts but need more holistic support that often goes beyond the current budget. What executives need, they say, are efforts that will strengthen their organisational cultures. They want data skills such as strategic planning, process improvement, increased cross-functional collaboration, operational excellence and an emphasis on critical thinking and problem-solving. In a way never previously seen in our industrial history, the need for ongoing data makes us all interdependent on other functions (human resources, operations, learning practices, etc.). By recruiting and managing these specific skills across all functions, an organisation can cultivate enthusiasm for data and a mature mindset for problem-solving, extending its initial efforts without extending its expenses.

Curating data as a service is much more than delivering monthly reports – data is used to drive every decision and uplift every function with more thoughtful decision-making. Data possibilities must shape strategy, which shapes the data team's priorities. In one study, 93 per cent of companies indicated that they plan to continue increasing data and analytics investments.[51]

As data leaders are expected to do more with less, upskilling is becoming more critical than ever – and it must start with leaders. Leadership skills, long thought of as 'soft' intuitive skills, and management, often seen as a 'hard' science, need to be replaced with a more hybrid skill set of 'informed intuition', where intuitive decisions are data-informed.[52]

The difference between ad hoc efforts and moving towards a sustained, intentional and curated data service comes down to a critical shift in the way we think about how we align our resources, how we partner and how we acknowledge driving business value *between individuals who understand how to, and are motivated to, integrate data into their day-to-day tasks*. A recent study about data skills for career progress states that 'the number of jobs requiring digital skills (including data skills) is predicted to increase 12 percent by 2024'.[53] The World Economic Forum's 2020 Jobs Report notes:[54]

- 24 per cent of employers think finding employees with the right skill set will remain their biggest challenge over the next five years.

- 50 per cent of all employees will need reskilling in the next five years.

- 85 per cent of Americans believe digital skills will be important to success in today's workplace.

- 94 per cent of business leaders expect employees to acquire new skills on the job.

---

50 https://www.forrester.com/blogs/15-01-28-digital_skills_are_the_golden_ticket_in_2015/

51 https://www.ey.com/en_us/ccb

52 https://www2.deloitte.com/us/en/pages/public-sector/articles/leadership-intuition-meets-the-future-of-work.html

53 https://www.urban.org/sites/default/files/publication/100843/foundational_digital_skills_for_career_progress_2.pdf

54 https://www.weforum.org/reports/the-future-of-jobs-report-2020

Strengthening leaders with data: an example

Decision-making: *Netflix blends complicated viewer analytics with years of experience exploring new product ideas.*[55] Starting with data, the final decision is made with informed intuition. In the case of Netflix, informed intuition works when reframing decisions between several organisational bets as an investment portfolio. From this perspective, leaders only need a smaller percentage of those risks to pay off big in order to compensate for failures. When making strategic decisions, hard data is only valid if placed in a highly human context, and data science might not be sophisticated enough to predict whether a product or solution will work among specific demographics. Ultimately, success rests on a bit of faith and lived experience as to how end users will react to a new idea. Applying informed intuition requires courage and conviction that the decision chosen is right, the willingness to be perceived as wrong for a time until the decision is proven right and the ability to articulate the logic behind the decision.

Your organisation faces a broad spectrum of challenges that call for a step to be taken beyond ad hoc data efforts, towards deeper engagement with data team members and stakeholders who can help you improve many areas of organisational resilience and development. We're not underestimating the value of ad hoc data projects – they can make a big difference in helping the business to move forward. However, investing in sustained efforts can address the same needs of the internal data community while building in-house talent, building bridges to the corporate sector and strengthening an organisation's ability to deal with constantly evolving social, economic and market challenges.

Table 2.2 contrasts ad hoc data efforts and ongoing data as a service.

Table 2.2 Ad hoc data efforts versus data as a service

Ad hoc data efforts	Data as a service
1. **Direct or immediate needs:** Headcount assigned to solve acute issues; extra hands. Minimal context is needed.	1. **Organisational or long-term needs:** Infrastructure teams and leadership require a consultative understanding of the business stakeholder context and needs.
2. **Type or support:** Hands-on learning on the job; significant lateral movement between analytics and engineering as most work stays at surface understanding.	2. **Type or support:** Deep skill expertise across data management functions; can think beyond the problem. Strong 'translator' skills between business and technical.
3. **Sample activities:** Regular reports, ad hoc reports, minimum platform skills/ investment.	3. **Sample activities:** High investment in platform (features, support, ongoing needs); regular insights and formal research plan.

---

55 https://www.forbes.com/sites/gregoryferenstein/2016/01/22/netflix-ceo-explains-why-gut-decisions-still-rule-in-the-era-of-big-data/

FIVE KEY IDEAS FOR MAKING DATA PROJECTS WORK

'Every successful organization has to make the transition from a world defined primarily by repetition to one primarily defined by change. This is the biggest transformation in the structure of how humans work together since the Agricultural Revolution.'

– Bill Drayton, Social entrepreneur[56]

This section is for data leaders or those who will use or have used data services in some way. Depending on your history with data projects, it might seem like this guide paints an overly rosy picture. You may be wary of investing more political capital, time, budget and the like. But the perception of a risky investment or diversion of limited resources, fails to recognise the ongoing enablement of data as an asset that data projects are meant to deliver.

In serving and researching digital transformations, there are five key ideas to realising an effort's full potential.

1. *Know and define needs clearly.* Take the time to define the project's data requirements. Data strategies are often organisation-wide and implemented through a series of constituent projects. This requires thinking through and communicating the context for the need. From there, engage data resources proactively according to your organisational priorities.

2. *Get the right resources for the right job.* Successful data projects align people, processes and scope to address specific decision-making needs; this includes adequate sponsorship.

3. *Be pragmatic (and ruthless) about deadlines.* Be thoughtful about prioritisation – even small tasks can take longer than expected, and data requests that require deep analytics or significant adjustments to data models or pipelines are rarely a good solution for urgent needs. Ruthlessly prioritise deadlines.

4. *Understand data as an asset.* If business stakeholders are to consume data accurately, ethically and to a certain standard, they must be involved in the process at each stage of the supply chain. (The same is true in reverse: if the data team wants to provision data accurately, ethically and to a certain standard, they must engage the business at each stage of the supply chain.) This means that context is critical to success.

5. *Learning goes both ways – always.* A data project is an ongoing partnership: the business supplies knowledge of its domain, while the data team brings deep ability (and a fresh perspective) to help translate business requirements into technical requirements.

Values and commitments, if not lived, represent only lip service and jargon. Table 2.3 looks at specific examples in which these five key ideas were either heeded or overlooked – with very different results.

---

56 https://www.forbes.com/sites/techonomy/2012/03/12/ashoka-chairman-bill-drayton-on-the-power-of-social-entrepr eneurship/?sh=2e2416c3698d

Table 2.3 Key ideas implemented and lived versus overlooked or ignored

One: Know and define needs clearly

Lived	Ignored
A team makes a reporting request to gain insights on a launched product feature. The product manager knew what they wanted the report to answer – to confirm or debunk a set of hypotheses for their feature – and the combined product and data team dedicated a large percentage of their group conversation to discussing this need. With a clearly defined end goal, the consultant team hit a home run on the project.	A product manager approached a peer group for a 'better report'. When pressed for details, they said they wanted this key document to look better. Look and feel were concerns, but what really troubled the team was their poor marketing content. No one on the team was bold enough to name the problem. They were concerned the feedback would be perceived as an insult. As a result, after months of work, the team got a very pretty year-end report that had the same messaging and even fewer insights. Owning up to the real problem could have resulted in a stronger data story that the team could use in various mediums well beyond the year-end report.

Two: Get the right resources for the right job

Lived	Ignored
An internal learning and development team collaborated with a group of data analysts and strategy planners to update their training process for new managers. The group used the deep expertise of its members, tapping design thinking, creative problem-solving and change management tactics to co-develop an innovative solution. This new perspective allowed the stakeholders to think entirely outside the box about what managers needed. With the help of a unique skill set, what initially began with the goal of updating existing training slides by moving the training to a more expensive location turned into a fundamental rethinking of the training procedure and a truly transformational impact on the organisation's new-manager programme.	Management was thrilled when the learning team brought new product managers through onboard training. Here was a way to systemically educate data-first principles across the function. They hired 62 new people in two months, almost doubling their headcount. But while the new hires were intelligent and promising, the onboarding process proved challenging on both sides. New employees learned language and frameworks for problem-solving and decision-making rigour that were not well mapped to job descriptions, performance management tools, day-to-day activities or annual goals. Managers were left to use their discretion on implementing key product management concepts, contributing to inconsistencies in performance and impact. The lack of reinforcement through existing mechanisms (such as performance management, manager 1:1s, functional business reviews, product methodology frameworks and the like) and inconsistent application by management caused unnecessary confusion and disappointment for new hires trying to live these new principles.

(Continued)

Table 2.3 (Continued)

Three: Be pragmatic about deadlines

Lived	Ignored
A business unit in a fast-growing organisation developed a new product and needed help preparing a go-to-market strategy in time for a major annual conference. Launching at the conference would likely result in three new high-profile customers signing. The plan wasn't due until the following quarter. Still, knowing the potential for a slow turnaround on feedback from a few senior stakeholders, the business leader began socialising the new strategy with an elite subset of the working group and key sponsors before seeking buy-in from the extended group. With a clear scope and need already in hand, the small but mighty group was able to deliver a draft of the strategy well ahead of the upcoming quarter's deadline, giving the business leader ample opportunity to refine their efforts among the broader working teams and sponsors – especially the more problematic sponsors – well before the annual conference. The product was featured in the keynote conference. The product was featured in the keynote and secured five new customers.	Excited about its latest ethics policies, a data governance group decided to engage a creative resource to update their materials for external viewing in time for an annual conference. The governance group rarely used design resources and wasn't aware of the creative group's priorities. When the assigned designer was pulled into a higher-priority project (also for the conference), the governance team had to point to an unvarnished text-only view of material at the conference, releasing the final designed content several weeks later. The experience confused external customers and partners and left the governance group frustrated.

(Continued)

Table 2.3 (Continued)

Four: Understand data as an asset

Lived	Ignored
The data team is lean but well-funded with a five-year strategy. It has taken steps to ensure that their inventory does not walk away. It tries to hire enough people with the right skills, place them in the right jobs and nurture leaders.	Due to a culture emphasising lateral career movement, the data team is understaffed and under-skilled in data management. Employees are functioning as data scientists without a data science background. Data engineers are not fluent in data management systems. Data resides across multiple systems, and there is a general understanding of where it is but no documentation. People are brought into the group based on their relationships and networks.
The data team obtains data correctly the first time. It makes it easy for stakeholders to find, access and understand, increasing trust and empowering stakeholders to use the data confidently. The data team protects its investment from being stolen or misused by adhering to industry standards.	The data team acquires data with little standardised intake, requiring significant cleansing efforts downstream. Stakeholders are confused by the multiple versions of reports and the lack of a single data dictionary. Not all data users can access the data they need to do their jobs. Trust in data is low since some stakeholders prefer to prepare their own data and believe their versions to be more accurate. Neither the data team nor the leaders consuming data adhere to basic risk, security or privacy standards, sharing files across teams and internal–external platforms.
The data team seeks to integrate data into the day-to-day running of the organisation, to enhance decision-making.	The data team feels frustrated by its inability to influence business decision-making more impactfully.

(Continued)

Table 2.3 (Continued)

Five: Learning goes both ways

Lived	Ignored
A business unit had made several ad hoc requests to its data team but needed deeper, faster insights before executive reviews. Because of the high level and manual nature of the reports the data team had provided to date, the data scientist could not deliver fast insights unless she was solely dedicated to that effort. After discussing the opportunity with her manager, she loaned herself to the business's 'fast insights' needs for two quarters, gathering insights on the executive's concerns. The data scientist gained first-hand exposure to the unique challenges presented at the executive level. The data team learned more accurate, efficient reporting requirements for supporting standard reports. The data scientist brought valuable knowledge from the executive perspective to the day-to-day business activities of the rest of the team. Each party came away from the experience with both individual and shared accomplishments.	A business unit partnered with an external consultant to guide a strategy session at an on-site meeting. The consultant had an excellent reputation for their human-centred processes and was also very busy, so they did not make enough time necessary to meet before the session to confirm the goals and expectations of the meeting. When the day arrived, the consultant took the group through a popular goal-setting process focused on making revenue targets. However, the group was primarily composed of internal functions (human resources) and focused on impact, not revenue. Because the consultant curtailed the full opportunity to learn (and because the stakeholder group moved ahead without these opportunities), the information was positioned entirely wrongly for the audience, and the training did not meet expectations.

GETTING STARTED

In the next chapter, you will learn how to identify and articulate the right data projects for your needs. But before you pick up a data project, it's best to determine your starting point. Two attributes – where you are as an organisation and your previous data experiences – will define how you approach future data projects. Let's take a moment to reflect on these two critical attributes.

Common roadblocks to success

Technology is important, but it's *people* that drive the change, implement solutions and sustain the work. According to Gartner in 2020, 'Data and analytics is not a technology implementation – it is a change management initiative.'[57] See Figure 2.1.

Figure 2.1 Most critical roadblocks (Source: Gartner, 5 Pitfalls to Avoid When Designing an Effective Data and Analytics Organization, Jorg Heizenberg, Alan D. Duncan, 1 May 2020. This Gartner report is archived and is included for historical context only.)

Most Critical Roadblocks

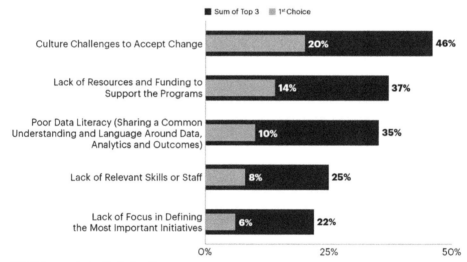

n = 291 All Respondents, Excluding Unsure
Q. Which of the following are the most important roadblocks to the success of your Data and Analytics team?
Source: Fifth Annual Chief Data Officer Survey

This book will help you mitigate these risks with helpful guidance in the coming chapters. In the next chapter, you'll review a step-by-step approach to scoping your data project. Scoping will help separate wants from needs. Phase Two: Determine Resources focuses on securing the resources and providing suggestions to find that 'next step' by blending project management and change management methods. Phase Three: Manage the Work

---

57 Gartner, 5 Pitfalls to Avoid When Designing an Effective Data and Analytics Organization, Jorg Heizenberg, Alan D. Duncan, 1 May 2020. This Gartner report is archived and is included for historical context only.

focuses on successfully managing the work by getting everyone on the same page, ensuring the project stays on track and ensuring the human–machine contributions are clear. Phase Four: Tune the Change suggests ways to fine-tune your efforts to manage data transformation in a way that strategically increases your capacity to do more. Use all the materials and be *hands-on* as you think about the best way to launch your data projects. Enrol your stakeholders as fellow revolutionaries – step towards a more data-informed culture by learning to assume your role in the data supply chain.

Your current organisational maturity can help define your data project needs

Too often, data teams view data maturity through the lens of asking 'What is our number?' and seek to prioritise interventions that only raise their team's maturity. Data maturity is not just a team measure; it's an organisational measure. Often, data teams can focus on the wrong things, making little difference. Once a number is obtained, it requires constant managing and fine-tuning to drive improvement, starting with the data team but extending through business stakeholders and, ultimately, the whole organisation.

Remember the projects from Chapter 1? There was the operations lead determining an executive scorecard, a programme manager implementing a CRM and another lead planning an SAP migration – to name a few. Often, these projects appear independent, but should be coordinated as part of a holistic data strategy. Too often, teams take their priorities from the issue backlogs of these projects instead of viewing them as interconnected. In this way, every data and technology project is an opportunity to think more strategically about how the organisation's data ecosystem can better align with broader business strategy. Alignment is lip service until teams can actually do it.

Not every organisation is ready for the same types of data project interventions. Organisational structures, revenue models and management approaches evolve as an organisation develops. Your 'next step' will look different depending on whether you are in start-up mode, fast scaling or a well-established enterprise looking to maintain and sustain its efforts.

In a period of rapid growth, for example, data services can be invaluable in building the foundational systems needed to move from managing ad hoc investments to managing rapid growth. In contrast, a renewal stage data project can bring much-needed perspective to an organisation that is perhaps getting too set in its ways – sparking the innovation necessary to rejuvenate a tired programme or team by offering deeper analytics and insights.

Table 2.4 illustrates how to use data projects to develop your organisation and the level of project management and type of projects likely at each stage of organisational growth.

What are your previous data experiences?

Most organisations have driven data projects of some kind or other, so the odds are that you have played one or more kinds of data roles. Your experiences might be pretty diverse, from good to bad to ugly – though I hope for more good than bad!

Table 2.4 Data project development stages

Formative phase	Rapid growth phase	Mature phase	Renewal phase
Teams or organisations lack a clear vision and direction when starting. Strong leadership and a willingness to experiment with programme design are keys to success. Infrastructure isn't a priority.	Teams or organisations move into a period of growth as they gain clarity, focus and momentum. With this growth comes a need for supporting infrastructure and relying on strong contributions beyond just the leadership team.	As growth slows, teams or organisations focus on incremental changes and improvements. Management moves into the spotlight, and leadership shifts out of focus. Infrastructure is deprioritised as systems are stable.	Recognising a need for transformation to address changes in the sector or operating environment, teams or organisations respond by adapting programmes accordingly. Leadership must balance the day-to-day needs with new ideas and work reminiscent of their earliest growth days.
Breakthroughs measure success.	Changes in the scale and scope of programmes define success.	Success is defined by the reliability and quality of services delivered.	Innovations in programme delivery measure success.
Target data project:	**Target data project:**	**Target data project:**	**Target data project:**
Focus on informal, organic data projects designed to build foundational systems that allow staff to focus on other priorities – including programme development and experimentation. At this stage, leadership and teams often work collaboratively – even blurring responsibilities – and much of this early data management support may come directly from senior leaders.	Focus on developing core sustainable systems that provide the infrastructure needed for quick expansion in staffing, programme delivery and making the business case for cost centre investments.	Integration of data analytics and data management across core functions should be the priority at this stage. It is also critical to apply data management to update the systems created in the early stages of development to reflect current needs. This typically manifests as vendor or resource consolidation (reducing redundant systems and processes). Finally, data management can address needs that may have been deprioritised during rapid growth to ensure no gaps in organisational needs.	Data management needs to echo those seen in the maturity phase. Still, data management can be used to redefine organisational development and support the reimagining of programme structure and goals. Outside perspectives can also help organisations think about the bigger picture and innovate.

Before moving to the next chapter, take a moment to reflect on your experience driving data projects to date. What do you want to preserve? What do you want to improve? Use Worksheet 2 to reflect upon what went well and poorly with past data projects and what you and your data partners and peers could have done differently.

With mature data management efforts, visions start top-down at the organisational level. Projects are defined to implement the vision. Visions can also be developed for specific projects. Projects in less mature organisations begin as an informal list of to-dos or backlogs. For these situations, anchoring the team's efforts into some larger vision is critical. A strategy cannot be managed from a backlog of data activities continually being reprioritised. At some point, the team has to adopt the dynamics of short-term (to-dos) and long-term (strategy).

You will want to add capabilities to drive constructive and disruptive innovations continuously. The aim is always to enhance the organisation's operational experience with emotional connections to end users (internal or external) who are new to your efforts. Codify this ambition in a set of target values and measurable behaviour characteristics, translating them into its corporate language and specific needs. Use Worksheet 1 to assemble a draft vision for your data project. Revisit progress against this vision every quarter.

KEY POINTS

- When business stakeholders provide requirements without considering the underlying architecture required to support them, they add significant drag to their operations. Data leaders at every level must take every opportunity to educate their stakeholders about *everyone's* responsibility to nurture the data supply chain that enables the organisation's competitive edge.

- The need to not just collaborate but *intentionally integrate data into business strategies* is essential to maintain the speed of business.

- Pragmatic optimism is seeing the limited potential of what can be achieved through ad hoc efforts, realising the exponential benefits of intentional interdependence with key stakeholders – and knowing how to achieve it!

- The debate over the phrase 'data-driven' versus 'data-informed' arose from attempts to mitigate the unintentional overreliance on data. The fundamental difference between being data-driven and being data-informed is that being data-driven lets the data guide the decision-making process.

- The need for ongoing data creates interdependence on other functions (human resources, operations, learning development and the like); recruiting and cultivating data awareness across all functions as strategic partners can cultivate enthusiasm for data and a mature mindset for problem-solving, and extend initial efforts without extending their expenses.

- Values and commitments, if not lived, simply represent lip service and jargon.

- Two attributes – where you are as an organisation and your previous data experiences – define how you approach future data projects.

3   SCOPE THE PROJECT

There is probably a long to-do list in your head of projects that would help you advance your organisation's mission. In Phase One, you'll learn to prioritise that list and determine which would be the best data projects to launch.

We will walk through the critical steps to developing the right project scope (as shown in Figure 3.1). First, we'll look at your organisational priorities. Then, we'll explore where these priorities sit in the data supply chain and address each need. Using some simple vetting tools, you'll arrive at a list of projects that are good matches for beginning to think of data as a service. With your vetted list in hand, you can prepare to share the opportunity externally by creating the first draft of a scope document and selecting your internal team.

Scoping projects is an iterative process. The right scope for a project is the first working step in a partnership between the data team and the business unit. It's unlikely that the scope document you can create in this exercise will be the final version you use in a project with potential stakeholders. At the end of Phase One, you'll have a couple of carefully selected projects and a first-take scope document for each, and you'll be ready to search for and approach potential stakeholder partners.

Figure 3.1 The scoping process

Assess your needs and identify your organisation's potential data needs.	Determine where needs sit in the data supply chain.	Vet projects using the four criteria of a successful data project.	Select your top one or two project choices.	Communicate project scope and select internal team. Remember, every project is an opportunity to support 'data as a movement'.

UNDERSTANDING DATA WANTS AND NEEDS

Fewer than one-third of all organisations embarking on data transformation misdiagnose their needs.[58] They focus on a symptom rather than a cause or name something they're familiar with rather than the thing that's really bugging them. For example, many organisations looking for uniform reports haven't considered their key performance indicators or organisation-wide goals as foundational for achieving any initiative.

Without fully managing the digital supply chain by defining, storing, managing and reporting common key performance indicators, *as a constant flow* of data versus static files from ad hoc efforts, every data project will be weaker than it should be. With the right business questions and corresponding data, organisations can use several delivery mechanisms, from individual development plans to monthly business reviews. They will know they are focusing on the correct numbers to align resources and drive their strategy. Clearly articulating your data needs is one of the most important steps in investing in data as an ongoing service.

The easiest way to define data needs is to use your to-do list – it's okay if it's full of ad hoc efforts – to identify initial data priorities. There are all kinds of specialised assessment tools to determine what projects you might take on, and they can be helpful when you are ready for them. In the meantime, starting with your to-do list focuses your attention on where both you and your organisation are. From there, confirm that the projects fit your high-level strategy.

The confirmation step is key. You don't want to use resources for something not aligned with longer-term goals. (Sometimes, the idea that data fixes everything leads people to make hasty decisions about how to apply it.) Data should only ever be an input to decision-making, not the sole answer.

When leaders need to move quickly, it's not always possible to slow down to get agreement on key ideas; they develop their unique understanding of things and act from there. This is often called 'being scrappy'. The goal is to *move fast* without *breaking things* – and it works until it doesn't.

Once leaders achieve early success and reliable momentum, scaling as efficiently as possible becomes the goal. Then, the definition of 'being scrappy' evolves into *doing more with less*. For example, aligning reporting efforts across multiple functions might be a goal. Each function needs a shared understanding and agreement on basic measures. A data team serving five functional leaders might have five different interpretations of 'customer', 'renewal' or 'adoption'.

A strategic plan can clarify and confirm that a data service supporting all five teams is a priority. If these teams continue to want to educate internally and externally about an initiative, (1) centralised reporting (2) from a single source of truth are critical mechanisms to consider. Starting with a to-do list helps you move quickly from needs to concrete projects. This chapter describes how to do a to-do list-based assessment.

---

58 https://www.mckinsey.com/capabilities/people-and-organizational-performance/our-insights/unlocking-success-in-digital-transformations

A more complex approach to identifying data needs is to translate strategic goals into operational planning. A strategy-based needs assessment uses the organisation's strategic plan to generate projects by asking, 'How will we accomplish this goal?' For example, to help nurture data among stakeholders, the project lead might want to build better training around reporting efforts, assisting stakeholders in taking insights and linking them to both tactics and strategy.

The strategy-based approach is essential when making a case for the data management talent in the overall resource strategy. We'll cover the strategy-based approach in more detail in Phase Four: Tune the Change.

USING TO-DO LISTS TO IDENTIFY PROJECT NEEDS

Start where you are. Using a to-do list to identify projects is assessment-based intuition. A new list of projects isn't necessary; determine what might work with the current talent and confirm that those projects are important to the goals.

Being assessment-based regarding the to-do list means always connecting the task under consideration (that is, streamlining reports, creating a data dictionary, starting to catalogue data, designing a self-serve reporting area, re-coding a platform from an old code base and so on) back to strategic or multiyear goals (that is, efficiency, cost savings, customer delight and so on). These goals don't need to be the project's origin, but using a series of 'why' questions can help connect the task to a strategic priority.

There is a chicken-and-egg relationship between working bottom-up and top-down. Both have impacts and occur with certain organisational conditions. When data teams gain traction, many projects are framed as 'a group of data projects' versus a company-wide strategic capability. The former doesn't usually attract the best talent or the highest budgets, whereas the latter does. The chance of success multiplies dramatically when C-suite leaders give their time and attention to building data capabilities and ensuring that data initiatives receive the talent levels and budgets they need. However, leaders won't often give attention to something that hasn't yet been proven, and most strategic initiatives started as a group of bottom-up projects. Therein lies the problem, but you now understand how to link bottom-up to-dos with top-down multiyear initiatives.

If you are working bottom-up, use Worksheet 3. Start with a project list, then describe the specific tactic, project or action each data project will help you accomplish. Then, working down the list, ask yourself, 'Why will this help us accomplish our departmental, organisational and strategic multiyear goals?' While not all data strategies are built bottom-up (that is, listing all the projects that should be done), defining the data projects is the first step in any strategy. In other words, it works both top-down and bottom-up.

Do you prefer to identify projects using your multiyear goals? See Level Two: Practitioner in Chapter 6 to learn more about strategy-based needs assessment. If you don't have a strategic plan or other way of articulating your goals, see Appendix A for an example of a multiyear goal-setting exercise that you can use to vet your project list.

It is important to remember that multiple tasks or actions may help you accomplish each departmental goal, organisational goal and strategic priority. Always answer as many of the 'why' questions as you can. The key is to trace the action of activity back to the reason why it is important to the organisation.

Interested in seeing how this exercise might look in real life? Table 3.1 shows an example project that can be traced back to a multiyear priority. It shows how a data activity directly impacts business goals. For a list of potential projects, refer to Appendix B.

COMMON NEEDS THAT DATA PROJECTS CAN MEET

Are you having trouble identifying specific tasks or activities that need to be accomplished? Or are you struggling to connect those tasks to a defined data transformation strategy? Often, thinking about the challenges you face within your organisation or specific departments can help generate a list of viable project opportunities. Table 3.2 lists some common data team challenges expressed as business needs and correlating solutions. Neither column is exhaustive, but they provide a sense of the sentiments expressed and the to-dos that can correlate.

Table 3.1 Tracking data projects to multiyear business stakeholder goals

	Data team goal(s): Update customer database	Business goal(s): Increase the lifetime value of the customer
WHAT is the specific task, process or deliverable you are hoping to produce?	Need to modify the database to track a common, unique identifier for all customers. (*because*)	Steward data by cataloguing high-priority metrics (customer, renewal, adoption). (*because*)
WHY is this important to meet your departmental or functional area goals?	It allows the customer record to be referenced in the central index without confusion or unintentional overwriting from other records (*because*)	Need to implement a new loyalty programme for customers based on tenure, usage, etc. (*because*)
WHY is this important to meet your organisational goals?	Support business stakeholder's goals for improving retention and renewals. (*because*)	This year, our goals include improving retention and renewals. (*because*)
WHY is this important to meet your multiyear strategic priorities?	To enable the business to increase the customer's lifetime value, we must overcome internal data challenges (incompleteness, accuracy, redundancy, latency) by investing in typical data architecture and shared efficiency processes.	The strategic plan states that we must increase the customer's lifetime value; this requires an ongoing investment in a centralised data strategy.

Table 3.2 Common needs met through data as a service

Common challenges expressed as business needs	Correlating data service solutions (Not exhaustive)
'We need to build stronger relationships with our stakeholders and maintain those relationships over time.'	• Develop communications strategy, newsletter, etc. • Establish monthly dialogue and quarterly requirements review
'We need to make it easier for our stakeholders to provide ongoing business/data requirements.'	• Conduct stakeholder needs assessment • Confirm stakeholder journey map
'We need to raise the visibility of our services across the organisation.'	• 1-pager Q&As, MythBusters, helpful lists, concept guides, training and resources
'We must design and deliver the practices, technology, tools, data, insights and processes that enable easy, equitable and personalised experiences for our employees and ensure scalability.'	• Develop data strategy and initial roadmap • Formalise a central hub for delivering (source of record) data and communications materials • Self-service reporting layer • Customer segmentation
'We need to partner with finance and functional leaders to reduce the time we spend on budgeting and forecasting.'	• Budgeting process design • Accounting process design • Organisational budget design and development • Security systems audit • Database customisation
'We ensure consistency, completeness and accuracy of our data; we need a single source of truth.'	• Clean and catalogue data • Develop data schemas and structures • Determine data storage solutions • Determine metadata criterion • Determine governance plan

Your short list of potential projects

Once you've considered several approaches to building your potential project ideas, list them in Table 3.3 in order of priority.

Table 3.3 Applied practice: list projects of interest – to-do list

1.	
2.	
3.	
4.	
5.	

Now you are ready to take this list and test each potential project against the criteria for successful data project engagement.

FOUR BEST PRACTICES FOR GREAT DATA PROJECTS

Consider these four best practices for successfully scoping data projects. We'll review each in more detail in the following pages. Ideally, you and your team will discuss the questions in this list for each of your projects of interest.

1. *Scope*: Can you clearly define the work that needs to be done? Do you feel confident that it won't change throughout the project?

2. *Urgency*: When does this project need to be done? What are the consequences of not hitting that deadline?

3. *Knowledge needed*: What knowledge about your discipline, sector and organisation will the data transformation experts need? Is the outcome worth providing that education?

4. *General readiness*: Are your team and leadership open to this project? Will they have time to implement the project deliverables? Do they have the time to be engaged and advocate for it? Are internal stakeholders or external partners on board?

Vet projects as carefully as if you were paying for each one individually so that you don't end up with projects you don't need or, worse, set up to fail. Vetting projects is at the core of key idea 4 for making data transformation work: understand data as an asset. If you take project selection seriously, you'll find out just how many items on your to-do list can be handled by the right talent, where you might need to collaborate with other teams or hire externally.

Scope

One of the most significant challenges in any project is the potential for the project goals to shift and for the related work to expand. Confirming the scope of work – what all parties agree the project will include and not include – is a good way to avoid this problem.

Key idea 1: know and define needs clearly suggests that not having a well-defined scope jeopardises the project's success from the outset. One self-aware leader said of a project with a vague scope, 'I didn't know what I wanted, and as a result, we never got there; it wasn't clear, and it prevented the team from getting traction.' In contrast, the project is on the right track if everyone involved can agree on a project outline and a list of project deliverables (such as documents, meetings or systems).

Considering how scope differs when moving from ad hoc investments to sustaining long-term services is important. With a long-term service investment, a change in scope is addressed by well-laid team agreements, service-level agreements and data contracts.

Organisations that want to maintain a fast pace must recognise that they can't maintain speed and remain siloed. They must engage in deeper cross-functional collaboration to share a common understanding of what data is necessary and how it is interpreted. Even with this goal in mind, scoping is reduced to a ceremonial activity when a data team is asked to absorb upstream change, resulting in 'scope creep' – for example, if an entire engineering roadmap is expected to change overnight to accommodate a new product plan based on a shifting marketing event. All changes – from the most radical to the mundane – must be accounted for in team contracts. The benefit of these contracts is not only transparency but also cross-team collaboration. For examples of team working agreements, data contracts and simple service-level agreements for internal data sharing, see Appendix C.

In this phase, when you're vetting a project idea, you don't need a fully scoped document – all you need is an elementary outline of the project scope. You need to clearly define objectives to make sure the project is worth your time and is at least feasible for a small data team. In most cases, you won't have all the critical information, such as the tasks, timeline, resources, workflow and so on (for example, you may not have the technological expertise to describe the details of setting up a data cataloguing project), so don't feel like you need to have nailed down everything that will happen and who will do what and when.

For example: does your strategic plan have a communications component to drive awareness? Is a website or training plan needed for stakeholders to self-serve their reports? Do you need about twice the amount of enablement content you currently have to help stakeholders think about using the data you are providing them? Do you need the latest bells and whistles on your website, or is your target stakeholder group not really at the cutting edge? These are the kinds of questions that you'll need to address in order to vet a project's feasibility and match with existing data management capabilities and resources. Later in this stage, you'll build out a more detailed scope document to use as the basis for project discussions with any potential data management partners.

A common mistake that data teams make is proposing a scope that is too large, asking for too much or delivering things that aren't required to meet the proposed need. Some projects tend to suffer from this more than others. Technology projects can have a wide range of options, leading to significant scope creep. When measures are readily available, there is a desire to report whether they are impactful. Any project that impacts company-wide resources can grow exponentially. Watch the scope definition carefully.

Resist the temptation to ask for everything on your to-do list. If you want a metadata platform strategy, don't design the scope to tackle ALL data – that's simply unrealistic. Instead, first prioritise the metrics that executives use (Tier I measures) and the supporting (Tier II and III) measures. That's a much more compelling and constructive request than trying to capture all current data.

Even if you can't reduce the scope of an ideal project significantly, you might be able to break it down into chunks so that it's more feasible for a particular time frame (that is, think computationally). Asking too much can often intimidate or overwhelm the person or committee that needs to approve this and other requests, leading to nothing at all, while being realistic about scope can get you a good part of the way to your goal.

Urgency

Some projects are tied to specific, concrete deadlines. Given the nature of data transformation work, delivering on a tight deadline can be challenging. Remember key idea 3: be pragmatic about deadlines. Projects will rarely get done as quickly as you initially think they should.

Given this criterion, let's look at projects that would be 'bad' for data transformation efforts to take on (important but too urgent to be fulfilled) versus those that would be 'good' for the data team to take on (important but with enough lead time for some slippage, as well as for thoroughness in execution and delivery).

BAD for the data team:

- Writing a business requirements document that is due in a week.
- Having a data leader facilitate a monthly review for the governance sponsorship committee with no notice.
- Accommodating unique instrumentation for standard metrics by multiple teams.

GOOD for the data team:

- Developing marketing requirements for success measures determining the impact of future annual events up to six months in advance.
- Determining data governance policies to ensure data and assets are properly used and managed consistently by stakeholders.
- Collaboratively developing instrumentation guidelines for product teams as required launch criteria.

Knowledge needed

How much does a data team need to learn – about their stakeholders' needs, the overall business and the sector at large – to be successful? All projects require learning, and that is a good thing. For data and business stakeholder teams alike, considering data as a service is new. Both groups have run ad hoc projects. It can be hard to tie tactics such as securing a data warehouse (where data had previously been stored and ignored) to strategic goals of designing an efficient architecture to save costs. Here, key idea 5: learning goes both ways, is important to remember. Learning about what the business stakeholder does, how they do it and what matters to them (and why) is an exciting piece of the puzzle for data teams to understand. It is also an excellent opportunity for the business to understand more about the data supply chain, how they impact it and how both teams provide tangible value to the organisation by curating a genuine corporate asset.

However, a project with too steep a learning curve will be challenging to complete. In general, you should carefully consider projects requiring deep machine learning or artificial intelligence knowledge. For example, the main difference between training data and testing data is that training data is the subset of the original data used to train the machine learning model, whereas testing data is used to check the model's accuracy. The training dataset is generally larger compared to the testing dataset. Machine learning projects will fail without in-depth knowledge of many aspects of mathematics and computer science, attention to detail in identifying inefficiencies to optimise the algorithm and high-quality testing and training data.

In Table 3.4, the upper right quadrant indicates high-impact projects for which little investment in training is necessary. These are the most desirable places to start. The upper left quadrant shows high-impact projects that require significant knowledge and training resources that are more general in their skills versus those with growing depth in their abilities. These might considerably impact the organisation – but read carefully, as the time and resource investment for training might be draining.

The lower-right quadrant is occupied by low-impact projects that are much less significant; even though they require minimum knowledge and minimum training, they will likely serve as distractions rather than bringing real value.

Finally, the lower-left quadrant of the model represents those projects that are both low-impact and require a large investment in training to impart significant knowledge. These projects are generally time wasters. Avoid them.

General readiness

All data management projects require sponsorship to support ongoing success, even if the sponsorship is just you. Internal partnership is also essential. Your partners and stakeholders are critical factors in key idea 2: get the right resources for the right job. You need a strong magnet of support for your efforts, just as much as you need the right talent for data roles on your team. Imagine the challenges you might face if internal partners and stakeholders lost enthusiasm, motivation and interest in assisting your project work. At the same time, you need the data team to be excited to be working

Table 3.4 Assessing impact versus investment: a four-quadrant model

	Significant knowledge required	Minimum knowledge required
High impact	**Tread carefully** • Large investment training for an entire stakeholder team in a standard methodology (change management, Agile, etc.) • Significant impact on your work Example: *Kicking off an enterprise-wide change effort*	**High leverage** • Low investment training for key stakeholders • Significant impact on your success Example: *Facilitation by HR or outside consultant of a guided strategic planning session*
Low impact	**Time suck** • Large investment in training data management talent • Low impact on your success Example: *Building out an excessive number of budgeting scenarios for the sake of comparison*	**Distractions** • Low investment training data management talent • Low impact on your success Example: *Adding another 25 measures to the executive report just because we can*

with partners. Remember that the two-way learning between data teams and business stakeholders can be one of the great sources of value in a project.

Later in this section, you'll define an initial project team for your selected data management project. When vetting projects, however, you need to consider two things:

• *Do critical data team members want this project as much as you?*

Sometimes we pick up projects because we think they are important or the right thing to do, but the work is unsponsored, unbudgeted or deemed unnecessary. You must secure official sponsorship (not just an email) to proceed.

• *Do critical data team members have the capacity to support the work?*

Everyone in the organisation might be bought into the work, but if the correct information can't be supplied, or timely feedback or delivery of solutions provided, the project will not succeed.

Answering yes to both questions means you have the right team for the right job. Use Table 3.5 to start to design a team structure. Remember that all the projects, large and small, require a project lead. Consider how much time that person will need to manage the project and what other projects might be on their plate.

Table 3.5 Applied practice: list of team members

Project	Who?	Capability?	Motivated and willing?	Notes
1	Data engineer			
	Person 1	Yes	Yes	
	Person 2	Partial	Yes	Superficial knowledge requires some training.
2	Programme director			
	Person 3	Yes	Yes	
3				

Are you having difficulty identifying the right resources for a particular project? This could be a red flag that you may not have enough staff or the right mix of capabilities to fulfil your mission.

Three high-potential data management projects

Which of your projects of interest passed all four vetting criteria for data projects (1. clearly defined scope; 2. important and urgent need; 3. that the education needed does not outweigh the benefits of the project results; 4. that sponsors and working members needed on the business side are ready and willing to participate)? In Table 3.6, list the one to three projects that pass the test and can be completed on time and within budget.

Table 3.6 Applied practice: list of high-potential data management projects

1	
2	
3	

Now that you've been introduced to the four criteria for scoping successful data projects, you can vet every potential project using these criteria. Worksheet 4 includes a set of project-vetting checklists to help you check potential projects against the four criteria.

SCOPING THE PROJECT

Once you choose a potential project, the following steps will clarify the initial scope and build your data team.

Why is there so much emphasis on scope, you may wonder. It is because if we can't be trusted to describe the problem, why should we be trusted to fix the problem?

Experience dictates that clear, realistic, collaboratively determined scope is one of the most important factors of successful project management. Both PRINCE2 and PMI project methodologies[59] highlight critical factors for a successful project.[60] Regardless of your methodology, these factors all point to a careful scoping process:

1. A clear description of what the project will create (deliverable).
2. Clarity as to what the deliverable requires (critical success factors).
3. Clarity as to what will make the deliverable successful (critical success criteria).
4. Proof the success factors were achieved (key performance indicators).
5. Feedback intervals are determined, and expectations for constructive feedback are clear.
6. Ongoing communications to reinforce the benefits of change and communicate success stories.

Considering these factors ensures transparency among all stakeholders (which increases trust), provides measures for success (which increases accountability and the likelihood of remaining within budget) and sets the project up to stay on budget with a realistic plan.

If you get scoping right, you'll likely still be getting the data project right at the end.

Defining an initial scope

Ultimately, creating the project scope should highlight resource dependencies – service providers (if you are outsourcing) or internal partners, staff members with functional expertise and executive sponsor(s). The project scope serves as the roadmap, clearly defining what constitutes a successful outcome and the path to getting there, so all stakeholders must agree.

While the project scope isn't finalised until the project team is confirmed, take the time up front to create the best possible project scope that outlines the work required in a clear, tangible and timely way, to the best of your knowledge. This will ensure that you are prepared to make a solid and compelling case when you submit funding requests and will boost your confidence when making your pitch.

---

59 PRINCE2 and PMP certifications are both globally recognised certifications. However, some areas favour one type of certification over the other. PRINCE2 certifications are more popular in Australia, Europe and the United Kingdom, but PMP certifications are more popular in Canada, the Middle East and the United States.

60 Project Central, Critical Success Factors https://www.projectcentral.com/blog/critical-success-factors/

At this stage you'll look for broad strokes, not too many details. Even if you've been thinking about an awareness plan (either external marketing or internal change management) long enough that you know precisely who you want to interview, what five competitors they should review for materials, what time the plan should cover, and so on, you might not want to add these details to your scope document just yet. The team and partners will want to be part of defining the scope – don't scare them off by thinking you've presented every variable on the project or, worse yet, cast yourself in the role of project manager. You'll want the team to take some control, and therefore ownership, of the project scope.

Who will help you develop the initial project scope? Ideally, a team *Lead* will manage the day-to-day responsibilities or will be accountable to the department to which the project most closely relates and will take primary responsibility for exploring and articulating the project scope. A *Sponsor* with functional expertise in a related area can also help translate the need into a clear scope of work.

What should a scope document include? Although you don't need all these items right now, a complete scope will typically describe the following:

- A summary of the problem you're trying to solve – demonstrating an understanding of its impact on the business.

- What the project will accomplish – a list of objectives or critical deliverables, the number of expected or acceptable reviews of revisions, and any associated meetings or training.

- What the project will not accomplish – a list of what is considered 'out of scope'.

- High-level success measures, risks and mitigations.

- When the project can be considered completed – a list of completion criteria.

- Logistics – the total expected length of the project and a list of required resources (including total costs and budget).

Right now, it's enough to take your best guess at each of these items. Even if you deprioritise this project or it gets put on hold, you will find this a helpful exercise as you continue to plan the project with other resources – or decide against it for good reasons you discover here.

Use the following steps to develop a clear scope template for each project. Record the answers in the blank scope template shown in Table 3.7.

Step one: in scope – what will this project accomplish? and why?

Describe the problem statement, activities, deliverables and specific 'chunks' of work which the individual or team will accomplish. Where possible, include exact estimates of the quantity or depth of activity at that step – for example, the number of use cases you hope the team will develop or the types of measures included in a dashboard.

Eventually, the scope document needs to include specifics. For example, a *well-defined* scope might describe discovery like this:

- The data strategy will unify and align the Acme product division's business units and horizontal functions on their highest-priority business questions. Questions will be mapped to defined measures and audited for cost. Measures will be organised in tiers: Tier I: executive level; Tier II: business-unit specific; Tier III: product level. Two designates from each stakeholder group (a sponsor and a manager) will meet regularly to set business requirement priorities for measuring instrumentation, acquisition, storage and reporting efforts.

Table 3.7 Scope template, example

Project name	Data Strategy for Cross-functional Division	Owner	Jane Doe
Status	Scoping		

Scoping statement:

The data strategy will unify and align five business units, finance and development (engineering, operations and IT) on high-priority business questions and Tier I measures used at the executive level. Two designates from each group (a sponsor and manager*) will meet regularly to determine business, architecture and data stack requirements. Ongoing stewardship of the strategy will evolve into ongoing governance activities (guidelines and policies for acquiring, storing and using Tier I measures).

*Subject matter experts will be identified as needed.

Problem statement (inc. business impact)	**Stakeholders**
Summary problem statement:	**End customer(s)/Personas**
Limited access, non-standardised delivery and low trust in x-organisation business measures. The result is a high cost to produce needed reporting to measure business model performance effectively.	Primary: Data leader, data team (data scientists, analysts, engineers), business unit leaders and managers, finance
Business impact:	Secondary: Extended stakeholders and end users (execs, directors, managers, analysts)
No centralised data management processes are in place that are fully sponsored and owned. The data team is newly supporting this programme at the executive level.	**Executive sponsor(s)**
	See Data Sponsor List
	Team members
	See Data Core Team Member List

(Continued)

Table 3.7 (Continued)

Core success metrics:	Deliverables/dates:
Determine and prioritise high-priority business questions; 100% engagement by business stakeholders	Scope draft, kick-off (Jan)
Determine mapping for Tier I measures by Q1	Define & confirm roles – responsibilities for all committee members (Feb)
Audit measure list for data cost projections by Q2	Determine current state observations among persona groups & initial priorities for a roadmap (Mar)
Produce an initial data dictionary to include necessary elements from partnering groups, establishing a single source of truth	Confirm next steps (Mar)
Confirmed business requirements document for data engineering team by FYxx	
Consumable data dictionary for FYxx	
Draft change process in place for FYxx reporting priorities	

		Scope boundaries	
Risks:	**Mitigations:**	**IN**	**OUT**
• Budget constraints and headcount bandwidth issues of SMEs within teams • Full alignment and involvement across governance working groups, engineering and finance	• Continue to drive visibility across stakeholder units and other teams through strong coalition efforts • Drive momentum through quick wins and complete transparency to show value	• Change control process • Feedback, comms schedule • Dictionary requirements • Monthly KPIs and operational health measures • Drive alignment between working teams	• Platform updates • CRM development • SDLC ownership • Instrumentation • International data requirements

At this stage, a *good* example of a section of an in-scope description looks like:

- The data strategy team will identify stakeholders in the Acme product division. They will meet regularly to review high-level business questions and determine related measures. Significant time will be dedicated to gaining agreement on standard measures and understanding the cost of instrumentation, storage and reporting on them regularly.

At this stage, a *bad* example of a section of an in-scope description looks like:

- The data strategy team will meet regularly to review and prioritise ad hoc data requirements.

Now you try it. Use a clean version of the scoping document in Worksheet 5.

Step two: out of scope – what will not be accomplished by this project?

What you choose not to do is as important as what you decide to do. Spell that out in clear, explicit terms. Things you do not want to do typically include activities or goals that are closely related to or may commonly be expected from a project of this type, but will not be delivered for *this* project.

At this stage, a *good* example of an out-of-scope section looks like:

- The team will target stakeholders of vertical and horizontal business groups. They will meet regularly to review high-level business measures to set guidelines and policy. The team will not dictate existing internal development work or processes.

At this stage, a *bad* example of an out-of-scope section looks like:

- The team and stakeholder group will establish a governance strategy.

Eventually, a *well-defined* out-of-scope section created with a data team might look like this:

- This governance strategy group will not cover platform updates, customer relations management (CRM) of any kind, ownership of the business unit's software development lifecycle (SDCL) process, engineering code instrumentation or international data requirements.

Step three: completion criteria – how will you know when the project is complete?

Identify a specific and clear endpoint indicating the project's completion. This section should also reference the right of all stakeholders to raise concerns, risks and ideas at any time throughout the project. Observe if this project is genuinely one-and-done or offers an opportunity to feed into sustainment efforts.

Eventually, your completion criteria created with a data team might look like this:

- The project is complete when the team lead has delivered the final report, dashboard, requirements document, or whatever your deliverable is, along with any awareness communication or training required for stakeholders to use, implement or benefit from the work.

- Alternatively, the project is complete when the team lead or sponsor has decided that the project should be terminated and has communicated this to the group. It is complete when the team lead has delivered the deliverable, written necessary change management communications and completed essential team and/or stakeholder training on the deliverable.

Step four: logistics – what timeline and resources are important to the project?

In this step, the ideal or mandatory timeline for the project and any resources that must be present or provided to complete the project successfully is described. For example, setting aside a budget to break contracts might be necessary if you intend to consolidate your vendors.

At this stage, a *good* example of describing necessary resources looks like:

- Full participation will be necessary for the project. As governance priorities arise, the working group will submit requests to the sponsoring group.

At this stage, a *bad* example of describing necessary resources looks like:

- The data team to secure the required funding to execute their priorities.

Eventually, a *well-defined* scope created with a data team might include detailed descriptions of necessary resources, such as:

- Each sponsor agrees to fully participate, including (1) meeting attendance, (2) 15 per cent of their manager's business goals dedicated to governance activities, and (3) 5 per cent allocation of their total budget towards priorities relating to a common data architecture for the Acme product division.

In Phase Four: Tune the Change, you will learn more about the importance of working group and sponsor group partnerships for scoping and managing projects. Setting this precedent in your earliest projects will be helpful as you move to expand your mission in the company.

BUILDING A DATA PROJECT TEAM

'Teamwork is neither 'good' nor 'desirable'. It is a fact. Wherever people work or play together, they do so as a team. Which team to use for what purpose is a crucial, difficult, risky decision that is even harder to unmake.'
                                    – Peter Drucker, management consultant, educator and author

When considering the resources to staff your data project team, it's also helpful to consider how you can leverage resources in other groups as an extension of yours. Consider the distinction between data projects and data initiatives.

For the purposes of this book, a *data project* is a project or portfolio in service of specific elements of a data stack. It is driven within your data team by people who are in your data organisation – internal resources. A data project is a temporary effort to create value through a unique product, service or process. For example, a project lead might determine a build or buy decision for a metadata platform that will be owned and maintained by the data team.

A *data initiative* is a data project where an organisation aims to introduce a new value or achieve a goal important to the organisation's business strategy. For example, if managing metadata is new to the data team and business stakeholder groups, the plan to support the metadata platform requires tight collaboration. Data cataloguing activities must have agreement on standard terms, calculations, data sources and so on. Another initiative example is the creation of a governance council, where budget, guidelines and policies are determined by stakeholders outside of the data team but led by the data team. Figure 3.2 shows a proposed data team and your thoughts on utilising internal and external resources.

Figure 3.2 Simple data team structure and roles (example)

Data strategy lead (internal)			• Set up a data governance council for executive direction • Set policy and guidelines for ongoing data usage
Master data management (MDM) core team (internal)		Master data management (MDM) lead (internal)	• Attain leadership commitment • Build team MDM workflow • Determine data steward roles
Data science domain expert (internal or external)	Data science domain expert (internal or external)	Data science domain expert (internal or external)	• Build expertise and stewardship • Enable best data management practices
IT enabling resources (internal or external)			• Determine data architecture • Adopt AI solutions • Integrate analytics models

LEADING DATA PROJECTS AND INITIATIVES

After you think about how you want to design your project team, you should consider how your team's efforts need to expand, collaborate and integrate with other groups and functions. What sorts of skills do you need to augment your team? How can your services extend and integrate more intentionally with the business units?

It's important to allocate enough resources to the data team or project and to be explicit about who is on the team and why. Assigning roles strategically and articulating a decision-making process can make the difference between success and failure in any project.

The following are key success factors for your data team:

- Clear role definition.
- Including team members who are motivated, engaged and actively participating.
- Establishing the terms of each person's participation from the beginning.
- Securing both sponsor and working member commitment by having them adopt specific performance goals; collaborating on language to ensure alignment.

Which skilled person on your data team would be a good fit for managing a particular aspect of your data strategy?

Think outside their standardised roles within the team – the best contact for an architecture project might be especially good at levelling up the conversation and educating others rather than someone you consider an engineering expert.

Big project teams are not necessarily better – a bigger group complicates scheduling. If you have too many people on the team, members might assume that someone else is taking responsibility for a matter or that something is not their job.

When considering possible team members for each data project, think carefully about the strengths each would bring to the data team, and try to balance experienced and junior staff members. Key idea 5 reminds us that learning goes both ways, and projects can be a great training ground for your team. Staffing your team with too many individuals who don't have the core skills for the project could force some team members to spend too much time doing basic training rather than the project's work.

What team archetypes are important to a data project? In my work supporting data management initiatives in both large and small organisations, I've found that taking the following roles into account increases the odds of a successful project:

- Day-to-day contact: Who will be the point person for the project?
- Decision maker: Where does the buck stop for critical decisions?
- Subject matter experts: Who has technical knowledge relevant to the task?
- Post-completion implementation and integration teams: Who will ensure the project deliverables are used and maintained?

You might update this list as you prepare for a project, but having a solid team (even if it starts as a networked team) to kick off your efforts will help communicate your readiness for the work to potential sponsors. Use Figure 3.3 to learn more about each archetype on your data team.

Consider when working with different types of cross-functional teams

Managing progress: The team lead is concerned with tying tactics to strategy and demonstrating progress. To do this effectively, it is critical to involve as many perspectives and stakeholders as possible to spot potential resistance early and ensure the success of the engagement.

Managing relationships: Tactics are important to track, but relationships are currency. It is also essential to keep sponsors and those enthusiastic about your project in the loop – both formally and informally.

Figure 3.3 Team responsibility archetypes

Day-to-day project manager	
Someone who manages cross-functional resources and tracks progress against the plan. They ensure everyone has what they need and provide consistent communication.	
Approver(s)	**Decision maker(s)**
Someone who provides feedback and input to the project direction and deliverables produced. Can be one or two people. *Related titles*: Can be from any department or function; often a senior functional lead or manager. Deep organisational and functional area knowledge is beneficial to the decision-making process. It is helpful to have an approver who has undergone a similar approach.	Individuals with ultimate executive decision-making power to provide affirmation or push back at key milestones and decision points in a project. Should be able to represent the viewpoint of key stakeholders. *Related titles*: Executive director, vice president, director of a function. Appropriate person is dependent on the impact and focus of the engagement.
Post-completion implementation or integration	**The specialists (SMEs)**
Individuals with primary responsibility for sustaining and implementing deliverables or resulting actions. *Related titles*: Can be from any function or level of seniority; sponsors and working group members are ideal candidates to assist in evangelism of the deliverable. Organisational knowledge is helpful.	Individuals with technical knowledge of the process or end impact of the engagement. Provide information while the consultants are collecting insights to support deliverable creation. *Related titles*: Individuals related to the function the data project is supporting. Knowledge of the organisation is helpful. Diverse perspectives are beneficial to provide multiple angles for consideration by the team lead.

A project lead can wear multiple hats, from day-to-day contact to decision maker and even subject matter expert. If you follow the intent of the responsibility – gathering a range of perspectives, ensuring stakeholders provide input before a deliverable is considered final and knowing how decisions get or should be made before a decision presents itself – you can create a far smaller team than that which has been described in Figure 3.2.

More complex projects, however, often require more resources. If you are working on something with a more extensive scope or a larger team of data management talent, you'll want to try to fill each role with an individual to ensure the project stays on track.

CREATING A ROLES AND RESPONSIBILITIES TEMPLATE

There are all kinds of roles and responsibilities templates to help identify who is responsible, accountable, consulted and informed (that is, RACI) on a project. Below is a simple version to help get you started; use it to draft a roles and responsibilities template using descriptions of stakeholder and data teams (see Table 3.8).

Table 3.8 Roles and responsibilities template

Stakeholder team role	Skills needed	Name
1. Day-to-day	1.	1.
2. Decision maker	2.	2.
3. Recommender	3.	3.
4. SME	4.	4.
5. SME	5.	5.
6. Other	6.	6.
Data team role	**Skills needed**	**Name**
1. Day-to-day	1.	1.
2. Decision maker	2.	2.
3. Recommender	3.	3.
4. SME	4.	4.
5. SME	5.	5.
6. Other	6.	6.

Together with the scope document, this document will provide a clear, tangible summary of your project goal and expectations. These two documents are critical in securing and prospecting for potential providers and guiding work between stakeholders and your team once the project gets going.

CONDUCT A PRE-IMPLEMENTATION REVIEW

Transformation work is hard. Data transformation work is more challenging. A 2018 McKinsey study found that 16 per cent of respondents say their organisations' digital transformations (which include data skills and projects) have successfully improved performance and equipped them to sustain changes in the long term. An additional 7 per cent say that performance improved but was not sustained.[61]

As you progress with your selected data projects, you can test the proposed scope by conducting a pre-implementation review. Unlike a post-implementation review (in which you write down what didn't work after the fact), a pre-implementation review serves as a kind of upfront risk assessment, helping you envision ahead of time intentions, ideal outcomes and what could go wrong.

This process can be as simple as asking a few questions. Assume everything goes wrong with your project. Why did it fail?

If you have more time, think about it in more detail. Consider the project a dress rehearsal for a performance. Go back to the four criteria for evaluation at the start of this phase. Are there any 'red flags' (serious deficits) in the four areas? Was the scope too complicated to clearly define? Was the deadline too aggressive? Did the sponsorship rely solely on one leader? Did the project require too many specialised resources? Was there too much flux in the organisation to proceed and maintain momentum (as with a merger or corporate integration)?

If you want to get at critical risks, you can conduct a pre-implementation review exercise with internal team members after you brief them on your project plan.

1. Start the exercise by informing everyone that the project has failed spectacularly.
2. Ask the team to spend a few minutes writing down every reason they can think of for future failure – especially the things they ordinarily wouldn't mention as potential problems for fear of being impolitic.

    Example: An executive compensation index was developed but ultimately caused more confusion and disagreement because the CEO unilaterally determined the underlying measures in a way that benefited sales. While the index was transparent, it did not reflect the values or intent of the overall stakeholder group and, therefore, lacked buy-in and complete support.

    Example: A data team member is working with a business stakeholder to translate business requirements into technical systems requirements but is pressured by the business unit's sponsor (and boss) to overhaul the supporting platform simultaneously. The data team member doesn't have the time, budget or skills to manage a platform upgrade. After waiting for a few weeks without a response to their request, the sponsor withdraws their support and opts to revert to their own home-grown efforts.

---

61 'Unlocking success in digital transformations', October 2018, McKinsey & Company, www.mckinsey.com. © 2023 McKinsey & Company.

3. Ask each team member to read one reason from their list. Have everyone state a failure in turn, and repeat until you've recorded all possible causes for failure.

4. Group similar risks together and rank them in severity. Group voting can help speed the discussion along.

5. Gather one last round of input for potential mitigations to these risks.

6. After the session is over, review the plan. Your list of risks and mitigations in the pre-implementation review creates a risk register for you to manage. Consider whether the scope should be adjusted to address any identified risks.

In addition to helping teams identify potential problems early, this exercise serves several team-building purposes:

- It can help reduce the 'damn the torpedoes' attitude of people who are overinvested in a project's outcome – those who shout 'full steam ahead!' despite known risks.

- Describing weaknesses that no one else has mentioned values team members' intelligence and experience, creating an opportunity to learn from each other.

- Making time for critique and participation before the project kicks off helps to increase buy-in from the team, stakeholders and sponsors. It helps to gain advocacy and reduce concerns.

- It primes the team to remain alert to early signs of trouble once the project truly kicks off.

This activity is not meant to undermine the project or drain your support; it's important to reiterate that every project has these risks, and by honestly assessing the challenges you face, your projects will have a higher success rate.

Use Table 3.9 to complete the pre-implementation review exercise for each of your selected projects.

Table 3.9 Applied practice: pre-implementation review exercise

Project	Risk	Mitigation
Business requirements gathering for the next round of reporting.	*Business stakeholders are unable to translate business problems into clear technical requirements.*	*Work with business stakeholders to gather enough context to help translate problems into requirements.*
	Data team cannot draw on their domain knowledge to help business leaders identify and prioritise their business problems, based on which will create the highest value when solved.	*Share domain knowledge both ways. The business needs a better understanding of the data supply chain and their role in it. The data team needs a greater understanding of the business domain and how to enable it better.*
	Stakeholders will set expectations that the data team cannot fulfil.	*Share this concern with the team and ask them to take the time to educate one another and raise all boats.*
	Stakeholders will receive reports that will not meet their needs and blame the technical team.	*Be transparent and let them know where you know you need to invest.*
1.		
2.		
3.		

KEY POINTS

- Scoping projects is an iterative process. The right scope for a project is the first working step in a partnership between the data team and the business unit.

- One-third of all organisations embarking on data transformation misdiagnose their needs because they focus on a symptom rather than a cause or name something familiar rather than address a tough challenge.

- The easiest way to define data needs is to start with a simple to-do list. From there, confirm that the project fits your high-level strategy.

- While not all data strategies are built bottom-up (that is, listing all the projects that should be done), the first step in any strategy is defining the data projects. In other words, it works both top-down and bottom-up.

- Consider four criteria for successfully scoping data projects: scope, urgency, knowledge needed and general readiness.

- Why is there so much emphasis on scope? It is because, if we can't be trusted to describe the problem, why should we be trusted to fix the problem?

- When considering possible team members for data projects, think carefully about the strengths each would bring to the data team, and try to balance experienced and junior staff members.

- When managing progress, the team lead is concerned with tying tactics to strategy and demonstrating progress. To do this effectively, it is critical to involve as many perspectives and stakeholders as possible to spot potential resistance early and ensure the success of the engagement.

- When managing relationships, tactics are important to track, but relationships are currency. It is important to keep sponsors and those enthusiastic about your project in the loop, both formally and informally.

- A pre-implementation review serves as an upfront risk assessment, helping you envision the intentions, ideal outcomes and what could go wrong ahead of time.

- Throughout the delivery of the project, there will be several opportunities to apply the Data Oath from Chapter 1. During the scoping process, we must ask ourselves *whether we should know* about the data we intend to collect – and we need to continue to ask that question throughout the delivery process.

4 DETERMINE RESOURCES

Until an organisation is willing to invest in its data capabilities, aligning data resources to answer complex business questions will be like riding a bicycle to chase a Formula One racer and never catching up. Effectively scoping projects involves establishing trust to eventually scale resources. While a single project manager can handle some initiatives, most data projects require a diverse team for execution. For instance, automating reports may need data scientists and data engineers, while data literacy and cultural initiatives involve collaboration with various partners across the organisation, utilising existing mechanisms from human resources and learning communities. Allocating or borrowing resources for a project contributes to its goals, provides necessary expertise and ensures that project outputs meet user needs.

Data teams must secure, develop and leverage adequate funding, resources and capabilities to meet the ongoing and growing demand for supplying data for business decision-making. This chapter will provide you with:

- a high-level understanding of how project management and change management methods enable transformative change, leading to the scaling of your work;
- key qualities to consider when building data teams;
- comprehensive insights into the roles you will need to partner with and guidance on how to engage them effectively.

Given the complexity of transformation work, it is essential to blend project and change management methods to foster and sustain the mindset and cultural shifts needed for lasting change.

RIGHT-SIZING AN APPROACH THAT WORKS FOR YOU

As mentioned in the Introduction, not everyone needs functional certification or a graduate school degree to execute data projects well. You do, however, need to forge a clear path forward – and a rigorous one.

Project and change management methodologies are two different disciplines; both are essential to data management projects. Project management focuses on the final delivery of what is in scope, on time, within or under budget and high quality (as defined by your team in your organisational culture). Change management is focused on the adoption, implementation and *integration of the changes by the people* in the organisation. For example, if you are driving a data strategy – which informs a common architecture, could impact budgeting for those involved and will culminate in a governance model to sustain your efforts – the strategy must be *adopted by all stakeholders* to become successful. Every data project must be fully adopted in a reasonable time frame and utilised by a large majority of your ideal target. Stakeholders must gain proficiency in the solution or tools to help achieve the future state.

Sounds like common sense. However, many projects fail because of the desire to deliver quickly; we skip past fundamentals such as confirming that those who must adopt a solution know its need and why it must happen now. Further, we ignore whether others have a desire to change. Suppose we assume these qualities and skip right to determining others' knowledge and ability to adopt the solution and provide training. In that case, we miss the opportunity to cultivate goodwill and enthusiasm and develop the ambassadorship needed to sustain change. *Support can never be assumed.*

Many activities are not exclusive to one or the other method. Still, they are common to the two (for example, managing risks, developing clear and realistic objectives, developing a solid business case and so on). Important distinctions show up in communications and training. For example, where project management focuses on the percent-complete and the risks of the change, change management focuses on the *why*, *why now* and *how* of the change. Figure 4.1 illustrates how these two methods complement one another.

Figure 4.1 Project management and change management

Project management activities	Change management activities
Initiate project →	
Scope project →	← Conduct readiness assessments and impact analysis
	← Identify and begin building sponsor coalition
	← Select and prepare a change management team
Plan project →	
Establish objectives →	← Select and address anticipated resistance
Document approach →	
Define team and budget requirements →	← Communicate why the change is happening (sponsors)
Design solution →	← Prepare and equip the working group and sponsor group
Benchmark and gather data →	← Continue communications and sponsorship activities
Generate ideas and select concepts →	← Launch group and hold office hours/provide coaching
Model solutions →	← Reinforce key messages (sponsors)
Document requirements →	← Continue communications and sponsorship activities
Develop solution →	
Evaluate alternatives →	← Identify training requirements and develop training
Architect solutions →	← Continue communications, sponsorship and coaching

For many, integrating two methodologies is challenging and can be cumbersome. Project and change management responsibilities can be separated between two people or groups. They can reside in a single resource (such as a project lead) or group (such as a data team) – it all depends on how your organisation is designed and how clear the roles are. This book adopts the project approach as the dominant method for Phase One: Scope the Project and Phase Two: Determine Resources, and shares tools, insights and featured culture shift callouts from change management in Phase Three: Manage the Work and Phase Four: Tune the Change.

Table 4.1 identifies and compares activities in the contexts of these methodologies. The list is not exhaustive but gives a sense of how the methods differ and how both are

Table 4.1 Comparison of project and change management activities

Tasks	Project management	Change management
Defining project scope	Software products and versions, data migrations and integrations, interfaces, training, out-of-scope elements	All the changes required by the organisation to achieve the benefits expected and who the change will impact
Requirements gathering and impact evaluation	Impact on **processes, software elements and people tasks**	Readiness assessments and impact analysis **on people and organisation**, processes included
Risk management (assess, impact, and mitigation)	Resources, budget, scope, **lack of user buy-in**, requirements, design and development re-work, customisations, data migration, UAT, regression testing, cutover, security, business continuity	Resources, budget, scope, **resistance**, design of the change re-work, impact on external stakeholders, market, engagement and morale, employee turnover
Communication management and strategy	Communications plan for stakeholders and team members **about delivery, tasks, deadlines, milestones, completion** • Identify stakeholders • Assess interest and influence • **Engage and influence stakeholders for support in the project** • Develop a communications management plan	Communications plan for stakeholders and team members **about the change happening, sponsorship, engagements, success** • Identify stakeholders • Assess influence • **Manage aspects of engagement, awareness, resistance, and personal commitment to help them in the project** • Develop a communications management plan
Training	To enable the use **of software, applications, hardware**	**To support the changes in the organisation is required knowledge about the change and the needed skills** (it may include training)

necessary to grasp the underlying issues that project management does not directly address – emphasis in **bold** to reflect against change management activities.

Pay attention to the subtle differences between the activities in these two columns and apply them to your circumstances. Recall the pent-up demand behind several ad hoc projects on the data team or the delivery of a significant data project such as automated and self-serve reporting. The business wants more reports; it needs different views of reports, and more frequent delivery. The data team delivers against these requirements, only to be met with more changes, and asks for the next round.

At some point, delivery begins to take longer and longer. Stakeholders become more educated through data addressing their initial questions, and their follow-up questions become more sophisticated. The data to address more nuanced concerns lies deeper in the data stack. It often requires more cleansing and preparation. Usually, the data isn't structured correctly, so the data team tries workarounds. All of these tasks are time-consuming. To operate at the speed of business, stakeholders start spinning up their data efforts, creating confusion between data sources. Definitions and calculations become misaligned. Pretty soon, all those reports you took on the responsibility to deliver, and are still providing, are not being used as often. The business has moved on, trying to extract more relevant data to meet its needs – and frustration mounts on both sides.

Whenever humans are involved, emotions are present. Data elicits emotions running the entire spectrum, from the hopeful enthusiasm of early partnership to the irritation of temporary solutions; from confusion about what will be delivered to dissatisfaction with and burnout from siloed efforts. The project manager benefits from leveraging change management techniques to continually assess the awareness, desire, knowledge, ability and reinforcement needed to keep stakeholders and the organisation informed, wherever they are on that emotional spectrum.[62]

To apply change management, the formal change must be identified and defined. Remember that bit about our inability to let go of old paradigms without a new paradigm and clear vision to jump to? The from–to movement of thinking past ad hoc projects to a more operationalised view of data management is ambidexterity in action. We feel the discomfort of becoming bi-modal: able to accommodate short-term ad hoc efforts while balancing larger, more impactful efforts. Figure 4.2 illustrates conceptually how the change process works.

When we lack awareness and understanding of how the data supply chain works, particularly our role in that context, we have difficulty understanding our role as an advocate. Effective and lasting culture change requires effective advocacy.

Recall the to-do list generated from Phase One: Scope the Project. Think of the data project you selected. Stakeholders generally want the benefits of concrete data projects such as streamlining operations to produce more frequent reporting, the efficiency

---

62 'ADKAR' is an acronym for the five outcomes an individual needs to achieve for a change to be successful: Awareness, Desire, Knowledge, Ability and Reinforcement. The model was developed nearly two decades ago by ProSci.

Figure 4.2 Change management process: inputs and outputs

Input

A change to how the business operates (project or initiative) that requires group members to do their jobs differently.

Phase 1: Preparing for change

Phase 2: Managing change

Phase 3: Reinforcing change

Output

Group members:

- adopt the change
- use the solution
- are proficient

of a single supporting architecture and a comprehensive data strategy. At the same time, *they don't always want to walk the journey with you* to accomplish those projects. There might be assumptions that the data team just goes off executing and returning to deliver. When that happens, critical business context is lost.

Awareness and understanding of our roles in the data revolution are the first two phases of change management. Change requires people to shift mindsets, behaviours and long-term working patterns. For example, streamlining reporting means business stakeholders can no longer have as many ad hoc requests accommodated. That might not seem like a big mindset or behaviour change – but when the change is put in place, stakeholders often need to be reminded that report frequency is more important than constant iteration. A singular architecture seems like a good idea to everyone when they are reviewing the to-be vision of the future – but when data migrations start or systems are retired, people's sense of ownership becomes impacted. Either way, the change is met with a feeling of loss.

People's emotions around change provide a helpful diagnostic for project leads. Emotions are a window to a person's tolerance for change. The project lead must observe those emotions and align an appropriate intervention throughout the project. The task list in Worksheet 6 can help you determine what preparation might be needed before Phase One: Scope the Project.

Use Worksheet 7 to evaluate the health of an initiative at the beginning, middle and end of a change initiative to identify key risks. Through repeated application over the lifecycle of the initiative, the model and assessment enable teams to evaluate progress and improve initiative health with informed interventions.

For a comprehensive understanding of how project management and change management capabilities can be integrated, see Worksheet 8. For a more high-level list of projects and their key deliverables, see Appendix B.

THE IMPORTANCE OF SPONSORSHIP

Not all leaders know how to be good sponsors. Sometimes, executive sponsors agree to sponsor an initiative without realising what it entails. Sometimes, leaders assume that 'a good project manager can just get it done'. Certainly, good work can happen with strong project leadership. However, sustainable change (for example, ongoing adoption, utilisation and proficiency) cannot occur with only a single project manager trying to do the right thing. Sustainable change requires constant collaboration and partnership between the project lead, sponsor and working group members.

The executive sponsor has a direct impact on whether projects meet objectives. Projects with highly effective executive sponsors meet or exceed goals more than three times as often as those with ineffective sponsors.[63] It helps when senior executives have more than a passing interest in the outcome of the work. They authorise change and send messages supporting it to broad stakeholder audiences, indicating why it is taking place and the risks involved in not changing.

For significant shifts in data culture, the Primary Executive Sponsor is typically a senior leader (Chief Data Officer, Vice President of a data function and so on). For smaller projects where data efforts are just starting, this person could be a senior manager (of a function or small data team).

Regardless, sponsors need the following traits:

- *Strong communication skills*: They are vocal, articulate, comfortable communicating with all levels in the organisation and able to provide awareness by explaining the reasons for the change.

- *Strong influencing and engagement skills*: They create passion and enthusiasm for the change and are engaged and involved every step of the way.

- *Visible and supportive*: They are willing to be the visible face of the change and provide the necessary resources to support it.

- *Approachable and available*: They make time for sponsorship, attend meetings and contribute as needed.

- *Recognised leader with sponsorship experience*: The sponsor is credible and recognised as a leader. The sponsor has experience, is capable and has a strong understanding of the their role.

Preparing for the change ahead is essential before taking up a substantive data project such as a data strategy encompassing many other functions. Use the questions from Exhibit 4.1 to interview the executive sponsor in order to understand better the kind of change they expect and the differences between today and the future state.

The answers to most of these questions will help paint a picture of the future state, set expectations and validate your initial scoping assumptions. They will also become the communication talking points and establish the foundation for training efforts. To apply

---

63 https://www.pmi.org/-/media/pmi/documents/public/pdf/learning/thought-leadership/pulse/executive-sponsor-engagement.pdf

Exhibit 4.1 Discovery questions for an executive sponsor

1.	Why are we making this change?
2.	What would happen if we did *not* change?
3.	What is the scope of the change?
4.	Who will feel the impact of the change the most? The least?
5.	How will work processes change?
6.	What will end users experience?
7.	How will this change make us more competitive as an organisation?
8.	How will technology change?
9.	How will this change impact financial performance?
10.	How will or should the culture and values of the organisation shift?
11.	How long will it take to implement this change?
12.	What barriers or obstacles do we foresee?

change management, the formal change must be an identified and defined change. Refer to this early discovery effort when thinking through a plan in Phase Three: Manage the Work, and when considering your mission and guiding principles in Phase Four: Tune the Change.

THE PROJECT CHANGE TRIANGLE AND ASSESSMENT

ProSci's PCT Model™ is a simple framework that shows the four critical aspects of any successful change effort: success, leadership/sponsorship, project management and change management.

Success

Success starts by identifying the purpose for the change, the project objectives and the organisational benefits. Leadership, project management and change management support those goals.

Leadership/sponsorship

ProSci's research studies confirm that active and visible sponsorship is the number one contributor to successful change. Executives and senior managers who authorise fund and charter change initiatives must also lead and sponsor these changes. They already make decisions related to strategy, resources and schedule. These same leaders must

Figure 4.3 ProSci PCT Model change management triangle (Source: ProSci, PCT and 'Project Change Triangle' are trademarks of ProSci Inc. Copyright © ProSci Inc. All rights reserved.)

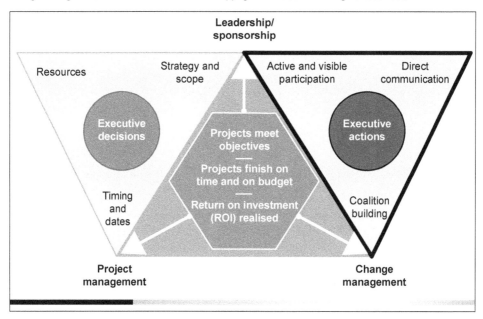

participate actively and visibly throughout the project, build coalitions of sponsorship and communicate directly with group members about the business need for the change (the *why*? And why *now*? concerns). Business leaders must also show how the change aligns with the vision and strategy of the organisation. The 'sponsor of change' role is difficult to delegate or re-assign because formal hierarchical authority and ultimate accountability cannot be trusted – no matter how flat the organisation is. Thus, one single project manager is never enough.

Project management

Project management is the set of processes and tools applied to business projects to develop and implement a change. One of the critical components of effective project management is having the change clearly defined – that is, *what is changing* (processes, systems, job roles, organisational structure and so on) – to manage that change effectively. Project management also involves understanding the trade-offs between time, cost and scope of a change. Finally, project management is a structured, disciplined approach to managing tasks, resources and budget to achieve a defined deliverable.

Change management

Change management is a method for a structured process and set of tools for leading the people side of a change to achieve a desired outcome. Change management must hold two simultaneous perspectives: an individual perspective (leading individuals through change) and an organisational one (how groups navigate a change process). Effective change management mitigates the risks of productivity loss, negative customer or end-user impact and group member turnover while maximising the speed of adoption and ultimate utilisation of the change throughout the organisation. Change management is the final element of the PCT Model and is essential for achieving the business results associated with a change and the focus of this chapter.

These interdependent teams work to meet project objectives on time, on budget and with the expected return on investment (ROI). Figure 4.3 shows the ProSci PCT Model, integrating the leadership, project management and change management concepts.

Use Worksheet 9 to better understand and assess four critical aspects of any successful change effort:

- *Success*: Clarifies the aim or purpose of the initiative.
- *Leadership/sponsorship*: Provides strategy, direction and guidance.
- *Project management*: Addresses the technical side of change by designing, developing and delivering the solution.
- *Change management*: Addresses the people side of change by enabling people to engage, adopt and use the solution to achieve results and outcomes.

Why consider these things?

Executive sponsorship is critical to an initiative's success, yet many project managers consistently report a lack of visible and active support from their executive business leaders. Nearly 50 per cent of teams rate the effectiveness of their sponsor as poor to fair.[64] The organisation might suffer from 'change saturation', making it very challenging for a leader to sponsor a high volume of initiatives and do them well. The leader might lack awareness of their sponsor role or the ability to execute it, thinking that if only their directs were stronger, the initiative would be a success.

The good news is that a strong project manager can impact two of those three challenges facing executives. The PCT Model (and related assessments) help drive a shared understanding and common language and context of what it takes to drive successful change. More importantly, it underscores the belief that change is a team effort.

Since every data project is an opportunity to seed a revolution, which by definition requires change, understanding change management's unique contribution relative to project management and leadership/sponsorship can help you take stock of the organisational climate and culture you are trying to shift. Mindsets matter.

Leveraging the model and assessment enables teams to identify organisational competencies and gaps by evaluating assessment results from multiple projects.

Lastly, change management activities might read like common sense. Many people skim through these tools and think they intuitively operate with these skills or operate from a mental checklist accounting for these things. Some might, but most do not. We think, 'Of course, sponsors should reinforce the behaviours they want to see.' But a shocking number of sponsors do not. We think, 'Of course, group members should understand the reasons for the change.' Yet, when asked, many will have conflicting interpretations of the reasons behind the new requirements or tools they now have to learn – to them, it's yet another ask, for which they have little time. For these reasons, the project lead must clarify roles and responsibilities to all members, integrate change management activities in the over-arching roadmap and confirm key deliverables for the project among all sponsors and group members. When done well, every data project can become a key lever in the data revolution that helps employees gain an edge in their careers, and the organisation can get ahead of the competition.

DESIGNING THE PROJECT TEAM

The reporting structure of the project lead and data team impacts what support, if any, the project lead can leverage for change management activities. The project lead will generally report through a leader aggregating data initiatives. Where data is becoming more centralised, the reporting structure is likely a Chief Data Officer. Where data is decentralised, the reporting structure could be through several potential IT, engineering, operations and finance leaders.

Again, it is important to underscore that mature efforts can happen in centralised and decentralised organisations. Maturity occurs when everyone interprets success and excellence similarly, regardless of the organisation's design. The concept of 'maturity' is a diagnostic of how clearly the organisation has defined the underlying infrastructure

---

64 https://www.prosci.com/blog/3-reasons-executives-fail-at-sponsorship

Figure 4.4 Preparing the project team

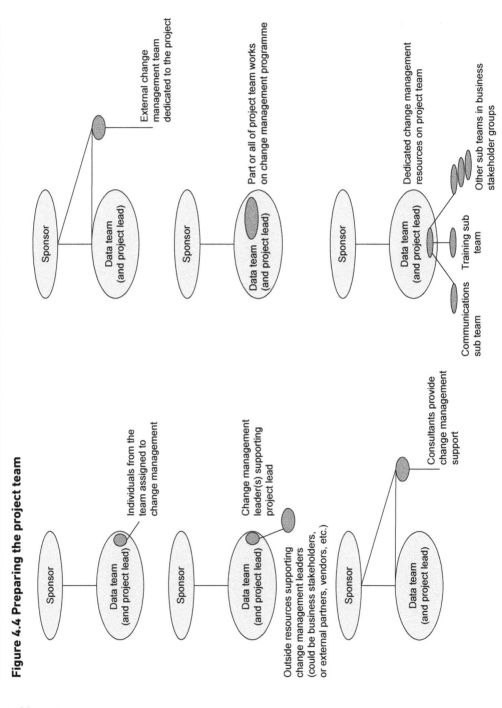

(for example, data stacks, data architecture and so on) that supports its ongoing reporting and analytics needs.

Finance, IT, operations and engineering can lead very successful data stack development in their functions and can also lead very successful cross-division/cross-organisation efforts. Some essential questions to focus on are: Is the data stack well defined? How are the stakeholders accountable for data usage? How broad is the end-user base of the defined data stack (within a single function, cross-functional, cross-organisational)? How well are those end-users served? Does the functional leader have *the right executive support* to continue serving those end-users? Is there a clear picture of how data is treated like an asset? The effort is indexing towards maturity if there are well-defined answers to these questions.

Team configurations

There are many ways to lead a data project initiative. As mentioned, sometimes the project leader handles the change and sometimes a separate change manager works with the project leader and the sponsor. Regardless of how the project and change management disciplines are integrated, it is important to understand how they relate to the executive sponsor to (1) leverage all three most effectively and (2) ensure all three disciplines are accountable for their roles and responsibilities. See Figure 4.4 for how the sponsor, project lead and change management roles can relate to one another in preparing the change management team.

Preparing the project team

We'll assume the project leader will take on the change management responsibilities for this guide. They might coordinate those tasks within the data team and/or leverage business stakeholders for some activities.

To ensure the success of the data project, the project leader must provide training to ensure consistent application of the data project's tools and change management talking points (for example, how data drives decision-making, how to extract value from data, how to use the tools, the importance of learning how to translate business problems into technical requirements and so on). You'll think through your general data training needs in Phase Four: Tune the Change.

The main thing to remember about communications and training change management activities is to always be driving the basic ADKAR principles of change management:[65]

- *Awareness* of the need for change.
- *Desire* to participate and support the change.
- *Knowledge* on how to change.
- *Ability* to implement desired skills and behaviours.
- *Reinforcement* to sustain the change.

---

65 ProSci's ADKAR model provides an excellent framework for many types of organisational change. It is recommended to utilise their change management framework and related tools as resources complimentary to this guide.

Preparing the sponsor model

Change complexity and organisational resistance increase together, causing a greater need for strong sponsorship. Figure 4.5 illustrates a variety of models through which sponsors can direct and support change.

But what does it look like to be a good sponsor?

Active and visible sponsorship is the number one contributor to successful change. Executive leaders and senior managers who authorise, fund and charter change initiatives must also *actively* lead and sponsor these changes. Referring to the ProSci PCT Model from Figure 4.3, three critical roles for primary sponsors are required for project success:

1. Participate actively and visibly throughout the project.

    - Allocate the necessary resources and funding.

    - Set expectations and establish clear objectives for the project.

    - Hold the team accountable for results.

    - Attend frequent project review meetings and actively review progress.

    - Remove roadblocks and make prompt decisions on project issues.

    - Be accessible to the project team; clear calendar when necessary to attend key events.

2. Build a coalition of sponsorship and manage resistance.

    - Build a strong sponsor coalition for the change among key business leaders and stakeholders.

    - Determine and communicate priorities between this change and other change projects.

    - Establish alignment around the overall business direction and the objectives of this change; resolve conflicting operational objectives with other senior leaders and middle management.

    - Ensure that managers communicate a consistent message about the change.

    - Recognise outstanding sponsors and manage resistance from those managers not supporting the change with their teams; enforce consequences for non-compliance.

3. Communicate directly with data teams and business stakeholders.

    - Build awareness with group members about why the change is necessary.

    - Share the risks or costs of not changing.

    - Show how this change aligns with the overall direction of the organisation.

    - Share the goals for this project and personal expectations with group members.

    - Celebrate successes with group members; be present and visible.

Figure 4.5 Preparing the sponsor team

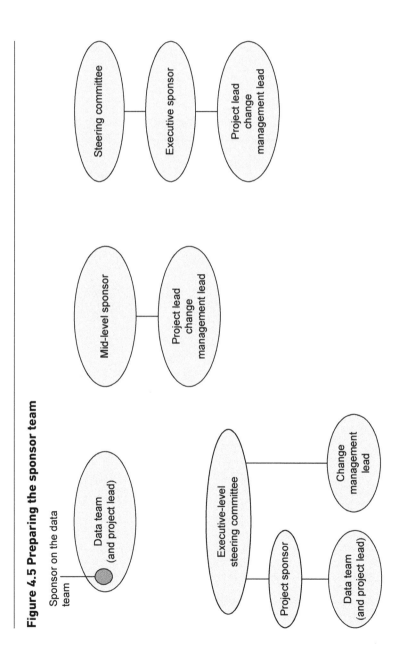

- Listen to group members and encourage feedback; be willing to answer the tough questions.
- Be willing to communicate to group members repeatedly to reinforce the message.

This list of behaviours can easily be used as a roles and responsibilities template you may discuss with your sponsor(s). See Worksheets 10 and 11 for sample behaviours you can use to gain consensus on mutual expectations.

Based on the Sponsor Model, use Table 4.2 to identify managers and business leaders to serve as project sponsors and working group members.

1. Identify the Primary Executive Sponsor (typically the leader that initiated the project)
2. Identify each group impacted by the change and identify who, from the group members' perspective, is 'in charge'. These managers are sponsors of the change.

Table 4.2 Applied practice: sponsor list

Impacted group	Name	Title	Role
Finance	Bob Smith	General Manager, Finance/IT Solutions	Executive Sponsor
IT, Solutions	John Doe	General Manager, Solutions	Sponsor
Engineering	Jane Mitchell	General Manager, Architecture	Sponsor

Sponsor diagram

Adding senior leaders between each sponsor and the Primary Sponsor listed above uses the hierarchical reporting structure. This structure shows the relationships between the sponsor and the Primary Sponsor. Leaders between the Primary Sponsor and individual sponsors are also sponsors. Add these leaders to the list of sponsors in Table 4.2.

Figure 4.6 is not an organisational chart. The diagram shows critical sponsors for the change (key business leaders must be on board for the transition to succeed). Sometimes the needed sponsors may not report to the Primary Sponsor, and several representatives may be from a single function.

Figure 4.6 Sponsor diagram (Example using sponsor initials in bold to indicate support and in default font to indicate neutral.)

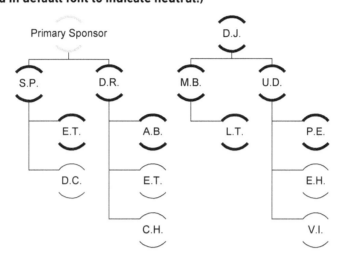

Assessing sponsor alignment to change

Having a way to assess the alignment of sponsors helps determine whether each sponsor identified supports, opposes or is neutral about the change. The assessment must be free of assumptions. When a sponsor's position is unknown, schedule a meeting to discuss the change and their position on it.

The project manager and the Primary Sponsor confirm the categorisation of sponsors. They determine if the members support (Group A), or, oppose or are neutral to change (Group B).

Determine the position of each sponsor regarding the change:

- Group A. Sponsors that openly support the change.
- Group B. Sponsors that oppose the change or are neutral towards the change.

Use Worksheet 10 to determine the current competency of each sponsor. If the project lead does not have enough information to complete the worksheet for a sponsor, request a meeting with the sponsor. Discuss these questions with the sponsor and ask them to assess their behaviours on current or past projects.

Assessing working group members' alignment to change

Now that the role of the sponsor is clear, let's look at how group members should be showing up. Understanding who supports, opposes or is neutral about the change is essential. Even informally administered assessments help clarify assumptions (on both sides). When a group member's position is unknown, schedule a meeting with them to discuss the change and their position on it.

The project leader confirms the categorisation of group members in the 'oppose the change or are neutral' Group B.

Determine the position of each group member regarding the change:

- Group member that openly supports the change.
- Group member that opposes the change or are neutral towards the change.

Use Worksheet 11 to determine the current competency of each group member. If the project lead does not have enough information to complete the worksheet for a group member, request a meeting with them. Discuss these questions with the group members and ask them to assess their behaviours on the current project or past projects.

Phase Three: Manage the Work explores the project method further and integrates some but not all of the concepts in this chapter. Every organisational culture is different. Some require more change management preparation, while others index more on the project management side. Review the material and generate the right balance for your needs.

Taking time to think through each one of these sponsor and group member behaviours helps the project lead to integrate planning deliverables more specific to change management with the project roadmap, such as:

- Communications plan.
- Sponsor roadmap.
- Coaching plan.
- Resistance management plan.
- Training plan.

Putting it all together: sponsorship in action

Clarity has now been offered regarding sponsor and group member roles and responsibilities. The project lead facilitates the Executive Steering Committee, the Working Group and stakeholder representatives in discussions ranging from prioritisation to feasibility, implementation, integration and, ultimately, sustainment. Figure 4.7 illustrates a simple stakeholder structure applying the principles of the PCT Model. The data project organisation consists of:

1. *Executive steering committee*: Primary prioritisation and escalation point for data projects that support the primary data stack.
2. *Data project/change manager*: Single point of contact; key facilitator of the data project as an ongoing solution (for example, strategy, governance, reporting/ analytics, experimentation and so on).
3. *Data working group*: One representative for each subject area is accountable for their respective functional area's engagement, implementation, integration and sustainment.

Applying data project organisation to a process flow helps to consider how the roles might work together. A simple process flow can stimulate discussion about how things currently work versus how they should work, presenting some opportunities to model more effective behaviours. Remember, every data project is an opportunity to plant

Figure 4.7 Example: data project stakeholder map

The key role in this recommended governance organisation is played by the Data Project Council, which consists of

(1) Data project/change manager – Single point of contact, project manager and key facilitator of the data project as an ongoing solution (e.g. strategy, governance, reporting/analytics, etc.)

(2) Data working group – One for each subject area, are accountable for engagement, implementation, and integration of their respective subject area

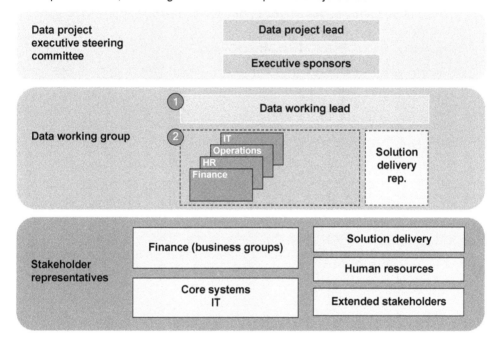

the seed for the data revolution – how you bring people together determines how the revolution will get started. Use sample data projects from Appendix B to review how the data project organisation will walk through how data projects are managed. Figure 4.8 illustrates the flow for how a new project request (for example, a request for a new metric or a new report) might be processed by the project lead, how its feasibility could be considered by a data resource, how it would get implemented by a data solutions team, and how it might eventually be prioritised by, escalated to or evaluated by the governing committees.

The diagram is where many business stakeholders and even some data team members can sometimes divide into two camps. Some find this structure reassuring. Others feel it is too much red tape.

Figure 4.8 Overview of data project workflow

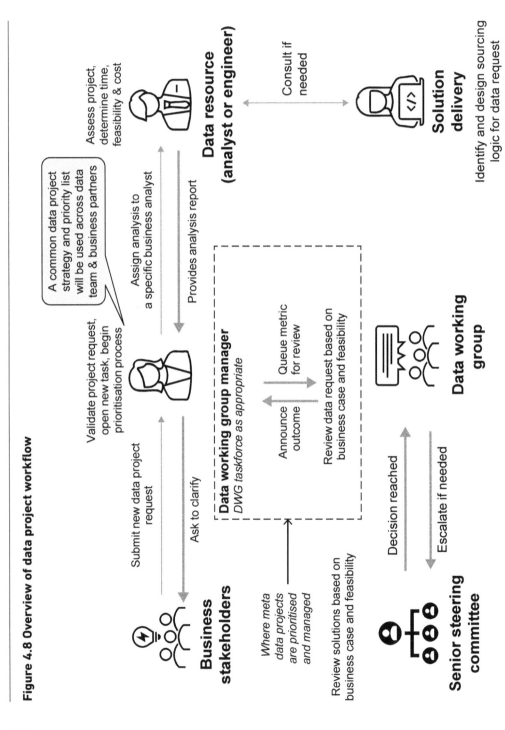

The thing to remember is that some aspect of this was already happening when requests were ad hoc. It just happened in reverse. An executive wanted to know a data point not accounted for on a report; a business stakeholder made a data request or asked a favour; a solution engineer or a data team member 'just made it happen'. The project lead was the last to know about it and, more often than not, added it to the regular reporting package as an ongoing delivery requirement – absorbing more workload.

> Tracking value drives accountability to everyone in the system and is the first step towards curating data as an asset.

The ad hoc method is viable. However, it will only ever produce 'data stuff'. It does *not* curate or produce data as an intentional corporate asset. *Why waste all that energy, enthusiasm, and talent on 'stuff'* versus pointing it towards cultivating an asset that adds value to the bottom line? The approach in Figure 4.8 still starts with the business's request, but the work's value is getting tracked. The request is validated to ensure it is addressing a key business decision versus a mere curiosity.

Know your audience

Different stakeholders have different, sometimes competing, drivers. An effective project leader must (1) know where their stakeholders are in their change journey, (2) share the correct information at the right time and (3) structure all communication around these drivers.

Every sponsor and stakeholder listens with two ears, filtering from two perspectives. While not exhaustive of all possible stakeholders, this split focus on the change at hand versus a functional agenda gives a sense of how different disciplines perceive priorities, needs and value. Exhibit 4.2 illustrates a steering committee meeting from a project and change methodology.

Exhibit 4.2 Two listening filters of stakeholders

Role	Project filter	Change filter
CDO, Executive sponsor	What is the status of the project? What progress can we highlight to show momentum towards our goals? What stakeholder can we highlight as moving in the right direction?	☐ Awareness ☐ Desire ☐ Knowledge ☐ Ability ☐ Reinforcement

(Continued)

141

Exhibit 4.2 (Continued)

Role	Project filter	Change filter
Finance Sponsor	Someone with a background in finance or accounting will look at the costs, numerical benefits attributed to a specific project and alignment with how they structure their data.	☐ Awareness ✓ Desire ☐ Knowledge ☐ Ability ☐ Reinforcement *Finance is on board with the initiative, except for one caveat: adherence to finance hierarchies. If we can't map financial, product and calendar year hierarchies (where not all using Gregorian calendar) into one, we can't proceed with structuring the data.*
Marketing Working Group member	Wants flexibility. The marketing team is agile by nature. Its day-to-day projects change depending on campaigns and initiatives. Want to make sure that the self-service reporting layer can adapt to change quickly based on frequently changing reporting needs.	☐ Awareness ✓ Desire ☐ Knowledge ☐ Ability ☐ Reinforcement *They have agreed to consume data from the central team for high-level measures but still run their reporting efforts. Their desire can sometimes waver, so their knowledge needs to remain current on what will be available and when.*
Sales Working Group member	Wants this data solution to make their job easier. Sales reps drive the company's growth and, as a result, have a heavy workload. They are most concerned about data projects that would have big changes for them; the ones that make their job easier are favourable, while those that make their job harder aren't favourable.	☐ Awareness ☐ Desire ✓ Knowledge ☐ Ability ☐ Reinforcement *Sales will benefit the most from this effort as the reporting layer will automate 50 per cent of their manual efforts. Desire is high, but there is concern over educating and training the sales group in a way that doesn't take them away from selling.*

Measuring success

As it relates to the implementation of the project, surveys can be used to track the achievement of results through the adoption, utilisation and proficiency of the solution and its supporting processes. Users will identify challenges in the current environment. Through quantitative and qualitative survey questions, seek to understand stakeholder perception of productivity, efficiency and effectiveness. Data collection is discussed in greater detail in Phase Three: Manage the Work.

Specific organisation performance metrics will also be collected to demonstrate achievement of the future state:

- *Speed of adoption*: How *quickly* stakeholders are up and running on the new solution, system, tools or processes.

- *Ultimate utilisation*: How *many* stakeholders of the total end-user population demonstrate buy-in and use the new solution, system, tools or processes?

- *Proficiency*: How *well* stakeholders and end users perform at the expected levels in the future state compared to what was expected in the design of the change.

DESIGNING THE DATA TEAM

Having explored the primary roles, responsibilities and stakeholder map for the project team (the sponsor, group members), it's time to examine how the project and data teams interact.

Four characteristics of effective project teams

- *Stable but flexible*: Employees are assigned to stay within a single team at the project level but can move between projects at the initiative level, depending on issue backlogs and context.

- *Lean*: Project teams should follow Jeff Bezos's 'two pizza' rule (about eight to ten people).[66] The smaller the team, the better the collaboration. Functional composition varies based on the project's needs and outcomes.

- *Matrixed*: Teams might be functionally (for example, engineering, programme management, data science and so on) aligned for skill development and performance but serve in cross-functional teams or communities of practice (for example, governance, data stewardship, business reviews and so on).

- *Focused management*: Project teams are managed by project managers (or programme managers). Leadership is generally flatter than the rest of the organisation.

---

[66] *'We try to create teams that are no larger than can be fed by two pizzas', said Bezos. 'We call that the two-pizza team rule.'* The smaller the team, the better the collaboration. Collaboration is also very important as software releases are moving faster than ever. https://docs.aws.amazon.com/whitepapers/latest/introduction-devops-aws/two-pizza-teams.html

Designing the data team for adoption, utilisation and proficiency

If every data project plants the seed of a data revolution, the data team is the incubator of adoption. There are three concepts to consider when designing data projects or initiatives for adoption (and scale): structure and leadership roles, competency and mindset.

Structure and leadership roles

To kick off the adoption of data projects and initiatives, the data team leader – the CDO or data leader – must define and fully resource the data team. The team should be structured to have autonomy. Even though the result is a service to be managed, the team should focus on achieving the end state (outcomes-oriented) versus elaborating on the step-by-step processes (process-oriented). As more business units and horizontal functions buy into a shared vision for architecture, reporting and governance efforts, data leaders will need to balance the types of roles on the data team, reducing intermediaries and investing in new leaders and deeper technical roles.

Data leaders should define data teams based on how data is consumed rather than how data is produced or the functionality it offers. For example, driving data insights is a work product, while an experimentation platform (that enables that activity) is not. Data project teams generate value by delivering capability or outcomes adopted by their intended internal or external stakeholders (or end users). They are oriented to an outcome or capability offered externally to customers, stakeholders or other functional teams. These project teams are called Scrum teams, Agile teams, hubs, pods or squads.

Data project teams start as small, execution-focused teams that generate value by delivering a capability or outcome that its intended internal or external stakeholders adopt. Often, these begin as ad hoc projects and grow into increasingly significant initiatives. Depending on the culture, some organisations want to keep efforts small, agile and short-term focused, while others require a shared perspective across functions. Reflect on Table 1.8 to think about embodying fluency from both perspectives.

The key point from that table was organisations clinging to an either/or culture to reduce their potential impact, and organisations embracing a 'both/and' approach to thinking and execution to generally increase their potential impact.

This is an example of how challenging it is to move between a 'from' and a 'to' vision and why change management is critical in helping to cultivate the mindset that helps stakeholders shift their thinking. The balance between centralisation, federation and decentralisation is unique in every organisation and representative of how it interprets (and lives) its organisational ambidexterity.[67]

Data teams typically sit inside technical functions (for example, finance, operations, IT, engineering and so on) that house interrelated capabilities. Those reporting lines often roll up into solution categories that coordinate roadmaps and share leadership resources.

---

67 This guide makes no recommendation on architecture implementation. In the centralised architecture, each institution must upload their data to a centralised web server, while in the federated architecture, data stay at their respective institutions but each institution must implement an interface to make the data findable though not necessarily accessible. https://www.researchgate.net/figure/Comparison-of-centralized-versus-federated-architectures-In-the-centralized_fig2_334205057#:~:text=In%20the%20centralized%20architecture%2C%20each,findable%20but%20not%20necessarily%20accessible

Competency

Once structure is determined, leaders must consider the skills and competencies needed to succeed. With the help of sponsors, the project lead helps prepare stakeholders for success, not just in the current state but for the future vision. Groups impacted by change need to develop new skills and competencies in the current state (such as the ability to translate business problems into technical requirements effectively) and the future state (learning to ask good questions – ones that drive the business forward). These skills emphasise strategic thinking, flexibility, digital fluency, cross-silo collaboration and business acumen.

Effective change management helps increase transparency on from–to shifts in stakeholder roles and responsibilities. For example, they may be getting new automated reports, which they will be expected to analyse and derive insights. Over time, a deeper partnership might emerge where they will use those insights to determine data science and research priorities for business questions requiring more specialised resources. The insights from this could impact future technical requirements. From there, the cycle starts again.

Competency expectations of group members (data team and business stakeholders) must be reframed by emphasising new skills needed for success, such as learning agility, digital fluency and collaboration. Sponsors can further emphasise a coordinated, cross-organisation approach to develop cross-silo upskilling by emphasising user experience and Scrum management. Centrally coordinating the development of new skills in the organisation (or existing skills used in a new way) helps transfer knowledge (and skills) to the rest of the team. For example, having project leads become Scrum masters through a central cross-silo collaboration initiative helps drive consistent organisational approaches and methods.

Communities of practice (CoPs) can take many forms. They can be domain-focused interactions, such as Scrum Masters from different Agile teams forming a CoP to exchange practices and experiences in building highly productive Agile teams. Role-based CoPs are one of the most common types of communities. For example, having key business stakeholders become data stewards through a central digital fluency initiative creates a single source of truth, ensuring consistent definition and application of data across the organisation. Informal networks are designed specifically for efficient knowledge-sharing and exploration across teams or the entire organisation.

The project lead can continue facilitating ongoing learning by developing communities of practice, coaching and mentoring where appropriate. As CoPs gain acceptance and participation, membership can be far more diverse and far-reaching.

Mindset

With structure and competencies defined, leaders must cultivate the mindsets and culture for success in this new model. Remember the human side of the data supply chain and how complex mental models are to shift. People are unlearning the old and learning the new. The in-between space of from–to is challenging without a clear picture of the end state. With sponsors' help, the project lead targets data initiatives that build members' mindsets of customer-first, agility, enablement, outcome orientation, silo-busting and cross-team collaboration.

It is important to note that project leads cannot succeed just by restructuring efforts and skill development. Mindsets, behaviours and ways of working must shift to ensure that members fully use the new model they work in – whatever data activity that might be. To succeed, group members need to move towards a mindset of becoming increasingly customer (or end-user)-centric; of agility and enablement, outcome orientation and cross-team collaboration.

The project lead must work with the executive sponsor (and other sponsors as appropriate) and be active and visible in helping impacted groups understand the mindset and behaviour shifts required by explicitly defining and socialising a 'True North' direction for the culture that other sponsors and working group members can understand. The project lead, in partnership with the human resource sponsor, should recommend adapting organisational competencies and performance metrics to drive behavioural change.

Measuring teams on business outcomes, end-user experience outcomes and their ability to collaborate across organisational silos help drive mindset and behaviour change. Cascading measures from the business initiative to project-level goals ensures that teams focus on enterprise objectives. For example, the enterprise goal might be to streamline and automate processes through operational excellence and system investment. At the data initiative level, the objective is to develop a single vision for data architecture. At the project level, this may translate to sunsetting specific systems, data migrations and federated access across multiple data silos. Each goal is tied to particular activities and targets at every level. To determine multiyear goals, see Appendix A.

KEY ROLES ON A DATA PROJECT TEAM

Progressive companies, ones that do *not* want to 'ride a bicycle to chase a Formula One racer', understand the importance of centralising data priorities and appropriately staffing their data teams. We gained some exposure to general data roles from Chapter 1, using real estate as a metaphor for understanding roles in a data organisation oriented to protect data as an asset. However, they sometimes struggle to define new roles that will support their efforts. Table 4.3 illustrates some key roles on a data project, their key concerns and core skills.

Organisational structures for these roles are illustrated in Figure 4.9. The combinations in which data teams may be organised are endless; every data leader will want to design their organisation according to the needs of the business. That said, design philosophies include centralised, decentralised or federated management approaches.

While there are still good reasons for highly centralised and decentralised data, most organisations opt for some form of hybrid. Federated data management can take a greater variety of forms than purely centralised or decentralised data, making them challenging to define, discuss and visualise. How a hospital interprets balance will look different than it will for a marketing company, yet both might consider themselves federated.

Table 4.3 Key roles on a data project team

Title	Answers these questions:	Description	Core skills
Project manager (also referred to as product manager or programme manager)	Is this initiative or feature important to the user? If so, how? Is it an important enough problem to solve in the organisation or product?	Defines the strategy, roadmap, requirements and releases of a product or solution. Determines and maximises the business value of the product or solution. Embodies the 'voice of the customer' and ensures the product or solution solves real-world problems and is fully utilised by the end-users.	IT leadership skills (end-to-end management of cross-functional digital initiatives) Experience creating strategies and roadmaps and considering business requirements Strong knowledge of continuous engineering/agile methods; ability to adapt Configuration management, identity access management, collaboration platforms, and other cloud-based services Financial management skills with operational budget tasks
Data analyst	What kind of data does the company have? How can we get the data? How do we interpret the results? How do we explain insights to multiple audiences?	Defines key metrics and builds dashboards illustrating business impact across multiple audiences.	SQL, Python or R Tableau, PowerBI or similar tools for building dashboards Statistics and storytelling How to run experiments Machine learning, such as regression analysis and time series modelling

(Continued)

Table 4.3 (Continued)

Title	Answers these questions:	Description	Core skills
Data engineer	What kind of and how much data does the company have? Where is it stored across the organisation? How secure is that storage? What is the operational health of that data?	Implements, manages and improves platforms and integrations with adjacent technologies to automate or reduce repeatable tasks and administrative overhead.	PowerShell, Bash and other scripting languages Management tools, techniques, monitoring and integration Configuration management, identity and access management Automation to drive improvements SQL
Governance lead	How much of the organisation's data is ours? Licensed from a third party? What decisions and processes can we automate? Where can we increase alignment between the business and data teams? How should we consider data access, sharing, storage and selling data?	Drives alignment between business needs and data architecture capabilities. Collaborates with engineering and operations to ensure coordinated delivery of data services – execution of data strategy. Develops and enforces data policies, guidelines, training and communications.	IT leadership skills (end-to-end management of cross-functional digital initiatives) Experience creating strategy, roadmaps and managing change deployments Strong knowledge of service desk processes Financial management skills with operational budget tasks
Change/ enablement lead	What is needed to adopt the product or solution? Does the end user have the motivation, permission and knowledge to utilise the product or solution fully? Do the learning goals for the product or solution align with any HR digital upskilling initiatives or other HR culture levers?	Defines enablement strategy, roadmap and processes for digital workplace enablement, product or solution. Develops, monitors and shares success measures about enablement impact. Works with shared service partners such as human resources, operations and communications.	Fluent in both change and project management methodology Experience leading technology change and deployments Organisational development Use case development; awareness of data tools, technologies and practices

Figure 4.9 Data team roles, organisational structures (Source: https://hal.inria.fr/hal-01765262/document)

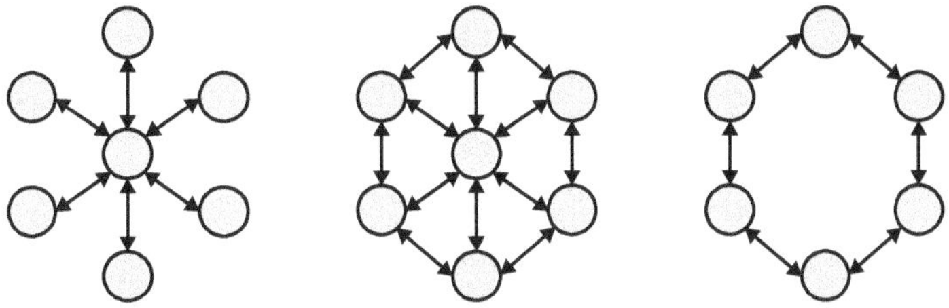

	a) Centralised	b) Federated	c) Decentralised
Culture	Leans towards a vertical style of coordination characterised by formal authority, standardisation and planning	Leans towards lateral coordination, characterised by meetings, task forces, coordinating roles, matrix structures and networks	Combines characteristics of centralised organisations (e.g. centralised planning, standardisation, etc.) and decentralised organisations (e.g. local leadership, competitive local objectives, etc.)
Common characteristics	• Decisions made by a single entity for the entire enterprise • Strong focus on policy setting and compliance • Strong central data org responsible end-to-end from data collection through to analytics • Lean analyst layer in business for ad hoc reports	• Decisions made locally but with support/ guidance from central governance hub • Local data groups conforming significantly to central standards mandated by a central team • IT data org distributed in each line of business	• Decisions made locally in the business unit • Local resolution of issues without escalation • No centralised power in data management (even if there is a central organisation providing some capabilities) • IT/data org distributed in each line of business
Key challenge	Can be challenging to implement and often requires a mandate	Inconsistent data needs to be standardised through communication/ coordination assistance	Leads to inconsistencies and difficulties consolidating at the enterprise level
Most effective when...	Single dominant 'business' type (e.g. retail)	Diversified lines of business Huge cross-domain priorities	No significant cross-domain requirements

Federated approaches typically have three main characteristics:

- A central governing body sets minimum standards.
- The central governing body provides a framework to aggregate and standardise data and make it available on behalf of the data owners.
- Use guidelines to confirm ownership and determine access to data. For example, executive-level measures could fall into a 'Tier I' governing category; secondary and tertiary supporting measures might fall into a 'Tier II' governing category; and business-specific data would be owned and managed in a 'Tier III' governing category.

DATA PROJECT APPROACH

Data projects help stakeholders understand the benefits of alignment – however that looks for your organisation. Alignment works best when it correlates to the amount of pain and frustration experienced in the business. A helpful way to think of solutioning is to 'be the aspirin, not the vitamin'. No one takes their vitamins regularly, but most of us will reach for aspirin when we are in pain. Too much or too little, and the change will not likely last. Constant calibration and fine-tuning will create sustainable shifts.

The good news is that the desire for data is high and only increasing. Organisations are starting to invest more in data operations, governance, data literacy and data culture roles to help enable lasting change.

Data projects help establish predictable delivery and ongoing change management of data and all the data-related programmes that support data as a utility and an asset. These programmes demand a mindset shift to achieve culture change. That starts by assembling *and aligning* technical data roles (data engineers, data scientists, data architects) and consumers of data (functional leaders and managers, casual business users and business subject matter experts) as consumers and stewards of the data supply chain. Alignment creates new roles, such as data stewards, requiring skills to translate business problems into technical requirements. Broader supporting roles such as data literacy and data culture tend to have broader use cases across the business, enabling everyone to learn to 'speak' data.

As mentioned earlier, the more complex the business question, the deeper a data engineer and data scientist must seek answers in the data supply chain – deep dives cause delays in data delivery. Data projects help bridge the gap between fragmented teams and siloed functions by establishing cross-functional programmes between data resources and data consumers, positioning stewards and data scientists as representatives within the lines of business supported by a central governing body.

Approaches: bottom-up *and* top-down

Organisations often feel they must choose either a bottom-up or top-down approach to data operations projects – neither of which extracts the full value of the data stack investment. Only a fluid *both-and* approach fully maximises data investments.

With the bottom-up approach, individual teams must cobble together the necessary data and technologies. This results in considerable duplication of efforts and a tangle of customised technology architectures that are costly to build, manage and maintain.

With a top-down strategy, a centralised team extracts, cleanses and aggregates data all together. This approach can eliminate some of the rework that occurs, but declaring victory with the initial cost savings undermines the value of managing data. These efforts are often not aligned with unique business use cases. When top-down efforts fail to support end users' specific needs, they are encouraged to spin up their own efforts, limiting data quality, time savings and cost.

Embracing *both* strategies will help lay the foundation for current and future use cases that will create value.

Thinking of data projects as data products can be helpful as products incorporate the wiring necessary for different business systems, such as digital apps or reporting systems, to naturally 'consume' data. Each type of business system has its own requirements for storing, processing and managing data. Consideration of how data is consumed in each aspect of the data supply chain (that is, consumption archetypes) is simpler and more agile than attempting to accommodate and prioritise hundreds of business use cases.

Data products built to support one or more of these consumption archetypes can easily be applied to multiple business use cases. The most common example of data consumption is self-service reporting; there are many other use cases. Table 4.4 describes five data supply chain consumption archetypes and examples of how they are consumed.

Table 4.4 Five data supply chain consumption archetypes (Source: McKinsey, https://www.mckinsey.com/capabilities/quantumblack/our-insights/how-to-unlock-the-full-value-of-data-manage-it-like-a-product)

Data product archetype	Requirements	Example uses
Digital applications	Specific data cleaned and stored in a particular format and frequency (e.g. delivering access in real-time to event streams or sensor data)	A marketing trends app
Advanced analytics systems	Data is cleaned and delivered at a certain frequency and engineered to allow processing by machine learning and AI systems	Simulation and optimisation engines
Reporting systems	Highly governed data with precise definitions – managed closely for quality, security and changes – aggregated at a basic level and delivered in an audited form	Operational or regulatory compliance dashboards

(Continued)

Table 4.4 (Continued)

Data product archetype	Requirements	Example uses
Discovery sandboxes	A combination of raw and aggregated data	Ad hoc analysis for exploring new business use cases
External data-sharing systems	Adherence to stringent policies and agreements about where the data sit and processes for managing and securing data	Monitoring systems that share fraud insights

Data projects enable:

- improved communication, and also help data teams, business stakeholders and sponsors adopt a standard worldview;

- improved understanding and awareness of the data supply chain that flows through the organisation;

- cost reduction of data operations through thoughtful consolidation, data pipeline and reporting automation;

- increased transparency, reliability and quality of data through monitoring;

- increased use and reuse of data, curating an evolving corporate asset.

Managing data as an ongoing service helps realise near-term value from the data investments *and* paves the way for quickly getting more value in the future.[68] Typical project success measures include, but are not limited to:

Money	Time	Quality	Quantity
• Sales	• Project length	• Meet quality standards	• Production
• Revenue	• Production time	• Defects	• Number of services
• Profit	• Down-time	• Code quality	• Service volume
• Discounts	• Task time	• Prevention relapse	• Service tickets
• Employee absenteeism	• First-to-market innovation	• Customer/ stakeholder satisfaction	• Customers
• Employee retention	• Approval time	• Meet environmental standards	• Inventories
• Cost saving	• Go-to-market time	• Positive media/ partner visibility	• Market share
		• Self-service rate	• Productivity (datasets are immediately usable)
			• Dataset access rate

---

68 https://www.altamira.ai/blog/what-is-data-as-a-service/

See how the project and change management measures blend together in Table 4.5. Specific success measures for a change project focus on engagement, development and adoption. The list below is just an example and not exhaustive.

Table 4.5 Success measures for project and change management (Source: Miller, S. (1997) 'Implementing strategic decisions: Four key success factors'. Organization Studies, 18. 577–602)

	DIMENSION	SAMPLE MEASURES
ACHIEVEMENT	**Did you achieve what was intended?** How well does what you implemented perform? Did you achieve your intended outcomes?	• Learning % staff trained / coached • Adoption % staff using new product, program, etc. • Performance % reduction in errors, % increase customer satisfaction • Vision: growth year on year, cost savings
COMPLETION	**Was the implementation completed?** Were all the intended aspects of the implementation completed within the anticipated timeframe?	• Schedule: scope, budget, planned/ actual • Completion of planned change supports • Effectiveness of change supports (e.g. communication, training, etc.)
ACCEPTABILITY	**Are the stakeholders and sponsors satisfied?** How happy are stakeholders with the implementation processes and outcomes? How enthusiastic are sponsors to sustain the change?	• Agreement with need for change • Satisfaction with the change product • Satisfaction with the change process • Trust in leadership • Perceptions of organisational change competency

Refer to Appendix D for a list of common data projects and their essential details.

KEY POINTS

- Scoping project opportunities *well* is about building enough trust to eventually scale resources. While a single project manager can accomplish some initiatives, most data projects require multi-disciplinary resources to execute.

- Every data project must be fully adopted in a reasonable time frame and utilised by a large majority of your ideal target. Stakeholders must gain the necessary solution proficiency or tools to help achieve the future state. A blend of project and change management methodologies should be considered for adoption.

- When we lack awareness and understanding of how the data supply chain works, particularly our role in that process, we have difficulty understanding our role as an advocate. Advocacy is required for effective and lasting culture change.

- Whenever humans are involved, emotions are present. The project manager benefits from leveraging change management techniques to continually assess the awareness, desire, knowledge, ability and reinforcement needed to inform stakeholders and the organisation wherever they are on that emotional spectrum.

- Not all leaders know how to be good sponsors.

- Sustainable change (for example, ongoing adoption, utilisation and proficiency) cannot occur with a single project manager trying to do the right thing, alone. Sustainable change requires constant collaboration and partnership between the project lead, sponsor and working group members.

- Sponsors must have strong communication, influencing and engagement skills; they must be visible and supportive, approachable and available; and they must be recognised leaders with sponsorship experience.

- Maturity occurs when everyone interprets success and excellence similarly, regardless of the organisation's design. The concept of 'maturity' is a diagnostic of how clearly the organisation has defined the underlying infrastructure (for example, data stacks, data architecture and so on) that supports its ongoing reporting and analytics needs.

- Regardless of how the project and change management disciplines are integrated, it is important to understand how they relate to the executive sponsor to (1) leverage all three most effectively and (2) ensure all three disciplines are accountable for their roles and responsibilities.

- The desire for data is high and only increasing. Organisations are starting to invest more in data operations, governance, data literacy and data culture roles to help enable lasting change.

- Alignment works best when it correlates to the amount of pain and frustration experienced in the business. A helpful way to think of solutioning is to 'be the aspirin, not the vitamin'. No one takes their vitamins regularly, but most of us will reach for aspirin when we are in pain.

5 MANAGE THE WORK

Now that you've identified your data project's resources, the focus shifts to efficiently managing the work. Utilising these resources effectively is just as crucial as securing them. After pinpointing enthusiastic individuals interested in solving data problems, the question arises: how do you guide them to achieve the right high-impact results?

Undoubtedly, steering data projects is challenging in any context. Leading a project or managing an ongoing process, especially one that represents a shift from the status quo, is inherently tough. Those working in data management are dedicated problem-solvers, committed to the mission and success of your initiative. However, having some tips and tools to simplify data projects is still beneficial. In Phase Three, you'll discover how the groundwork laid in Phases One and Two contributes to executing the project.

On the journey to greatness, you'll explore:

- how to effectively apply your scope document;

- what elements to include in key meetings;

- how to navigate your role;

- the phases and tasks to completing a data project;

- the best ways to support the data team, sponsors and stakeholders.

Most examples in this book reflect what happens in a specific type of data project engagement: a team-based project with stakeholders recruited from across the organisation, including the use of or partnership with outside partners. In the first half of 2020, Big Tech companies complied with 85 per cent of government requests for user information.[69] Companies such as Apple, Google, Facebook and Microsoft received more than 112,000 data requests from local, state and federal officials, with Facebook accounting for the largest number of disclosures.[70] When consumers get vocal,

---

69 https://www.newsweek.com/big-tech-complied-85-government-requests-handed-over-data-first-half-2020-1603070

70 https://apnews.com/article/how-big-tech-created-data-treasure-trove-for-police-e8a664c7814cc6dd560ba0e0c43 5bf90

companies sidestep their ire by developing new features, as was the case with Facebook declaring new platform encryption to protect users from future broad requests for data.

The point? The 'can we/should we' debate is complicated. There is no correct, obvious answer that companies and consumers can live with. Therefore, we must find the most ethical path to manage paradoxes and dilemmas. How should data be shared? Who has ultimate ownership of identity information? How should data be stored?

These examples and questions illustrate how the demand for data by third parties directly drives the debate and standards for consumer trust, privacy and security governing companies across every sector around the world. These incidents reflect the reality of data operations projects, where the primary goal is to keep multiple people from different teams across different functions and backgrounds aligned and working towards a single outcome.

Building foundational skills for managing easy or intermediate projects prepares you to transition to more advanced data work, such as weighing data privacy with data democratisation initiatives or thinking more deeply about the ethics of data storage or the sharing or sale of that data to third parties. On a scale from 1 to 10, where 1 is 'simple' and 10 is 'complex', where would you put your project or initiative? It takes a conversation among several people to get a feel for the complexity of a project or initiative. So, gather various perspectives and use Table 5.1 as a checklist.

Table 5.1 Determining complexity of projects

Simple	Complex
Few parts	Many moving parts
It's technical	It's adaptive *and* technical
Linear and predictable	Difficult to manage and predict
Fix what is broken (a part)	Work on and in the (whole) system
Magic bullet – quick fix	Integrated/systemic solution
One person can have the answer	Diverse perspectives needed
Can be solved and resolved	Must be managed, ongoing
Formulaic 'off the shelf' solution	Curiosity; experiment and innovate

Most mistakes happen because a problem was oversimplified or where a problem is solved independently by one department or function and applied to all. Looking at complexity solely through the lens of just one function or discipline creates a blind spot.

A frequent theme throughout this book is the need for intentional collaboration – a point that cannot be overstated. Engaging with complexity requires breaking down silos and effective, cross-functional collaboration between disciplines.

If there are high confidence levels for a specific direction, ask: *Could we be oversimplifying things?* Before answering, pause to reflect on the factors that make a strategic initiative complex and, therefore, more challenging to manage.

DETERMINING COMPLEXITY OF PROJECTS

There is a difference between complicated and complex.[71] For example, a puzzle may be complicated and have many separate parts, but a Rubik's Cube has many parts that interact with each other. Moving one part impacts all the others.

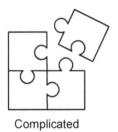

Complicated

Complex

Complexity requires considering the whole (however that whole is defined). No stakeholder, risk or data-related work stream can be viewed in isolation. Remember the list of data projects from Chapter 1? Many technology projects are, in fact, data projects.

As leaders increasingly depend on data to drive decision-making, business and data teams must come together more readily to craft their strategies, recognising their interdependence. Marketing, human resources, operations and other departments can no longer afford to work in isolation from data teams. Data teams have traditionally aligned well with finance for Sarbanes–Oxley compliance, have an even tighter relationship with finance, human resources and sales, and develop ethics, security and privacy standards that work for everyone – not just the most recent customer. All the organisation's parts must work well together, developing data translation skills. Most ambitious projects are not complicated puzzles but a complex Rubik's cube.

The danger with complexity is that it is often hidden, resulting in oversimplification, overconfidence and false certainty. That explains why so many projects are overly optimistic regarding timelines or budget, why most transformational projects fail and so on.

As you consider what data projects to move forward with, there are five principles to consider that contribute to success. These principles are not unique to data work, but they take on particular importance in this context:

1. **Prepare to sustain your effort from the start**. Data projects are rarely one-and-done. Track the projects you have done or want to do, and plan for what's next.

---

71 Kurtz and Snowden, 'The Cynefin Framework', in *The New Dynamics of Strategy* (IBM Systems Journal 2003).

A data dictionary was completed; how will it be updated and governed to stay relevant? How should user experience become involved to increase its usability? A data warehouse has been secured; how will data usage be monitored among stakeholders so you can better manage costs (that is, avoid ignore-and-store surprises)? KPIs have been determined; how will they be monitored to ensure relevancy as the business evolves?

2. **Embody your stakeholders**. Check in regularly, hold the team accountable for meeting deadlines, and give and receive feedback. You'll get better results.

3. **Communicate, communicate, communicate**! Keep internal team members and external stakeholders up to date, even if they aren't involved in the day-to-day project decisions. Without buy-in from everyone, even the best projects fail. (Remember the Rubik's cube? There is often more interdependency than we realise.)

4. **Engage in two-way learning, always**. Create the space for stakeholders to share feedback; you'll want to take full advantage of everyone's perspective and experience.

5. **Recognise goodness – be behaviourally specific**. Identifying behaviours accompanying outcomes helps light the path for *how* we want to travel together towards the goal. Behaviour is what (specifically) someone does to influence others, build support and achieve. From there, we mimic, creating a constructive sub-culture. Ultimately, we all want to do well and be seen as doing well.

Practical guiding principles, which we explore below in Figure 5.1, can apply to every data project you lead. They can help guide team conversations over the can we/should we debate that inevitably comes up when cross-functional data groups come together to collaborate over controversial infrastructure implementations or business models.

Figure 5.1 Five guiding principles of successful data teams

Principle 1	Principle 2	Principle 3	Principle 4	Principle 5
• Prepare to sustain your effort.	• Embody your stakeholders.	• Communicate, communicate, communicate!	• Engage in two way learning.	• Recognise behaviourally specific goodness.

With those principles in mind, the steps to starting a wide array of data projects are consistent. Table 5.2 lays out the key phases of any successful data project.

Recalling the PCT Model from Chapter 4, we are now focused on the project management side of the data revolution.

The rest of this phase describes each work phase in detail. What will you need to do and watch out for to make each phase of your project successful?

Table 5.2 Overview of key phases of a data project

0. Plan	1. Kick-off	2. Discovery	3. Build	4. Implement	5. Evaluate and acknowledge
• Set up effective working relationships • Review expected start and completion dates • Conduct general project orientation and other pre-work	• Bring internal team and external stakeholders together • Define and align the work plan • Review the end goal and outcomes of the project	• Team lead gains insights from relevant stakeholders and sponsors • Gather information from relevant sources (interviews, surveys, competitive analysis, etc.) to inform the development of deliverable(s)	• Lead begins first draft of deliverable(s) • Sponsors and stakeholders review deliverable(s) and provide feedback • Lead provides training to ensure successful implementation of deliverable(s)	• Lead presents final draft deliverable(s) and implementation plan (or other plan for use) • Lead provides resources (communications, documentation, training, etc.) to ensure successful implementation of deliverable(s)	• Acknowledgement of the success of the project and the team • Sponsors acknowledge achievements • Evaluation (including post-perspective) to update processes and outcomes

0. PLAN

Starting a data project requires preparation and planning. The goals in this phase are to clarify the primary objectives; to establish a good rapport with the team, the stakeholders and the sponsors; and to ensure that they are familiar with your team's mission and that there is reciprocal alignment with the data team's goals. Specifically, does the data strategy support the business strategy? Ensuring early support paves the way for a smooth kick-off.

Take time to establish expectations and review the scope, particularly the deliverables and the timeline. Invest time upfront to consider what might be needed to sustain your efforts once you've completed the project (Principle 1). Think about empathising with your stakeholder's needs (Principle 2) and how what you intend to deliver will help them. Learn from each other (Principle 4) by sharing context from your discipline and business perspective. Lastly, it is helpful to adopt the stance of not doing all the work yourself, but directing it. Whether managing a data project or a data organisation, this stance can be challenging for project managers and C-suite leaders alike. Genuine two-way learning (Principle 4) is more complicated than it sounds when you believe you have the answers.

In this section, we'll cover:

- Working with others.
- Understanding your role as a leader.
- Preparing to work with stakeholders.
- Keeping everyone motivated throughout the project.
- 'Preparing for kick-off' checklist.

Working with ... anyone

All successful projects (and organisations) are built on solid project management practices. Working with members inside the data team and with outside stakeholders (that is, partners, vendors or service providers) requires awareness (and *genuine empathy*) about their unique challenges. It provides an opportunity to blend your expertise with their knowledge and apply it to collaborate on a solution that will work for *both* of you. Fostering the kind of relationship that makes lives more manageable and ensures great work requires establishing trust, cultivating mutual receptivity and providing value.

Trust is critical to freely exchanging ideas, suggesting alternatives and supplying critical and positive feedback. Without confidence and respect, one party may not believe in the solutions, reactions or suggestions the other provides. The result? Good ideas might not be adopted or considered adequately due to a lack of trust in the other party.

Reciprocity ensures that solutions are developed collaboratively towards mutual benefit. If problem-solving is one-sided and a leader isn't receptive to input from stakeholders, partners or vendors, their perspective and expertise won't contribute optimally to the end result. Similarly, an overly assertive stakeholder may hinder the creation of an open and balanced dialogue necessary for fostering the best solutions.

Delivering value in data projects and transformations often involves cost savings, efficiency gains and improved data service health, all aimed at facilitating faster and more accurate decision-making. While data work is often unseen by the broader business, individuals within data organisations find personal satisfaction in achieving operational milestones. Fulfilling the need to 'do great work' and create a tangible impact on the organisation's bottom line makes data work more meaningful.

However, data project work presents unique challenges. Many business units may not have intentionally collaborated with data teams, leading to potential risks when the engineering roadmap undergoes significant changes. It's crucial to consider the ideal roles and responsibilities of the data team and stakeholder groups at different phases of a data project. Table 5.3 outlines the key qualities of data team and stakeholder roles, balancing desire/willingness with specific experience and knowledge.

Understanding role(s)

We are used to relying on our resourcefulness to solve problems. We find a feasible and sustainable solution when we are creative and proactive. However we arrive at solutions, we're also likely driving them. In this sense, working with partners or securing internal collaborations might require a shift in how we approach problem-solving. We are tasked with solving a problem *and* guiding the development of the solution. Look at Table 5.4 for an explanation of key differences between the role of a data project leader and a stakeholder.

The data project lead's role is to *amplify the exchange or value for their stakeholders.* Suggestions for recognition opportunities and ways to keep stakeholders motivated are highlighted in the subsequent parts of this book – but never miss a chance to connect the work to the end impact you expect to see. *When this is done with authenticity, it's like handing peers and stakeholders a tiny bonus.*

Table 5.3 Teams, stakeholders, partners and contractors comparison

	Motivation	Knowledge	Experience	Commitment
Data team	• Long-term commitment to organisation • Primarily motivated by mission, career goals • Require ongoing development investment	• Deep (horizontal) knowledge of mission, vision • Require deep context of the business need to become effective	• Have deep knowledge of organisational constraints • Grounded in practicality and reality	• Manage multiple responsibilities; often focused on immediate needs and not long-term fixes
Business stakeholders	• Long-term commitment to organisation • Motivated by synergistic goals – outcomes • (might) Require development investment	• Deep functional (vertical) knowledge of mission and vision • Require some awareness of data supply chain and how to 'speak' data	• Have shared knowledge of organisational constraints • Have a functional view of reality, lacks full perspective • Varied backgrounds bring new perspectives and ways of thinking	• Working with multiple stakeholders can mean multiple priorities
External partners	• Motivated by marketing/ PR or other synergistic expression of reciprocity	• Limited knowledge of mission and vision	• Have limited knowledge of organisational constraints • Operate with different values and reality • Varied backgrounds bring new perspectives and ways of thinking	• Working with multiple partners can mean multiple priorities
Contractors	• Require compensation	• Minimal knowledge of mission and vision; focused on discrete tasks	• Accustomed to working on budget-driven, time-limited engagements • Have benefited from exposure to multiple organisations	• If work extends past the contract, formal extensions need to be considered

Table 5.4 Leader versus stakeholder responsibilities

Leader responsibilities	Mutual goals	Stakeholder responsibilities
Leverage knowledge about the organisation and personal perspective on the challenge at hand	• Develop solutions to the situation or problem at hand • Build individual, team and organisational capacity and capability to implement *and sustain* the solution	Leverage knowledge about data management, technical subject matter expertise, and perspective from prior collaborations
Leader actions		**Stakeholder actions**
React, direct, respond, identify, give/solicit input, qualify, educate, drive awareness, inform, develop, track, monitor, observe		Research, discover, learn, create, drive, infer, suggest, present, consult, add perspective, provide context, solve, participate, contribute time/resources/ budget

The importance of balancing leading and guiding with doing in stakeholder management

Q: Why does it matter who does the work?

A: Effective stakeholder involvement is a fundamental principle underlying all data operations projects. A successful project is a collaboration with widespread support and involvement from the stakeholders identified. As the recipient of your services, it can be challenging to push the group as you would a team of direct reports. Stakeholders have their own interests and can block progress. They each have different expectations and different information needs. Too much gets lost in translation if you step in on their behalf. You must be clear about the stance you are taking when you consult, inform and engage participation to move the project forward:

• *Consultation* is a two-way process that includes the stakeholders in decision-making and planning. Stakeholders will provide information, opinions and ideas directly affecting the project's direction. This is a low-burden form of engagement.

• *Informing* stakeholders is more of a one-way communication of project-related decisions, progress and status.

• *Participation* involves direct contribution and involvement of stakeholders in the project that will influence ongoing ownership and long-term sustainability of the project outcomes, undoubtedly impacting budget and resource allocation.

Q: What is the harm in stepping in to do the work of stakeholders?

A: Stakeholder engagement is crucial to a project's success. This makes it one of a project's most significant risks. If you are a project manager leading a data project, you manage a group or a set of entities to accomplish a goal. If you are a leader of a data project or organisation, you need to focus on influencing, motivating and enabling others to contribute towards organisational success. Both data project managers and functional leaders gain more acceptance of their decisions when they communicate with multiple stakeholders and consider their feedback. This helps develop a strong sense of trust and commitment to the organisation.

Sound like common sense?

How often have you been tempted to step in and *do*, versus waiting for consensus on a decision or a deliverable? It seems most manageable for others when it is time to slow down and check the group's desire for change. When desire wavers (and it does, for many reasons), it is imperative to listen for what has changed and adjust accordingly.

Three common assumptions about business–data partnerships

1. **Data requests are one-and-done.** Even for an ad hoc request, all layers of the data supply chain need to be considered: acquisition, storage, aggregation, analysis, usage, sharing and removal. For example, a business unit might need to analyse a significant amount of data, and so request access to a data warehouse to store it. But a simple request for analysis impacts long-term storage costs. Was it a one-time analysis? How long should the data stay in the data warehouse (and who pays the rent there)? Too often data teams accommodate requests hoping to gain more context and insight into the business. Business stakeholders knowingly or unknowingly take advantage of data storage options without considering the costs. All this contributes to the myth that data requests are essentially free.

    Meanwhile, the data team is absorbing preparation and provisioning work along with exponential costs across multiple stakeholders over an indefinite period without understanding the impact this data has on the business. Many companies generate reports and maintain them in this fashion over decades. The sunk costs of data maintenance are no joke.

2. **Business stakeholders hold all the context necessary for a request.** A business stakeholder needs data to influence a decision. They present their request to the data team to deliver. The exchange goes something like this:

    Business stakeholder: *'I need x data by y date.'*

    Data team representative, usually a data scientist: *'What is the question you're trying to answer? What action or decision will you take with this data?'*

    With several follow-up questions:

    - *What is the impact of this question, and how will it help the company?*

    - *Who will be using this data?*

- *What time frames are crucial here (e.g. monthly, weekly, daily)?*
- *What is the visualisation you are trying to create?*
- *What interactions/drill-downs are required (type of use, revenue amount, etc.)?*
- *Are there any other details we should know about this data request?*

This type of back and forth can appear like the parties are two ships passing in the night – that people just aren't listening enough to hear one another. Sometimes, the business stakeholder hasn't thought through, doesn't know or isn't held accountable for answering these questions. Regardless, those three potential explanations can tap into defensiveness, irritation or frustration with what they perceive to be a simple request.

The data team representative is searching for context. They know that with the data request's context, a business is less likely to provide wrong or incomplete data requests, reducing the risk of making bad business decisions – a valuable point which data teams need to promote better.

3. **Data projects are self-funding.** The use of data can indeed increase efficiencies, but managing data isn't cheap. Many organisations are unaware of how much they spend on data because costs are diffused across the enterprise – ad hoc use of data masks actual usage and costs.

Additionally, costs for data management might be spread across many units. Third-party data costs might come out of the business unit's budget, for example, while reporting cost resides in relevant corporate functions and might manage data-architecture spending. Simply put, the total cost for a holistic data strategy that aligns and supports multiple business strategies is jarring, to put it mildly.

For example, a mid-sized institution with $5 billion in operating costs spends more than $250 million on data across third-party data sourcing, architecture, governance and consumption (Exhibit 5.1). How data cost breaks down across these four areas of spending can vary across industries. In smaller organisations, these operations can be even more opaque as different groups absorb all sorts of data activities. For example, industries such as consumer packaged goods that don't directly engage customers often have a higher relative spend on data sourcing. However, the result remains the same: managing data is a significant cost source for most organisations.

These are some of the most common assumptions about collaborations between business stakeholders and data teams.

What should business stakeholders know about working with data teams?

Most business stakeholders do not have experience with or awareness of the data supply chain. For many, the activities of the data supply chain will be new. Even if they are advanced data users, there are many things to take for granted about data and the infrastructure and resources that support it. The way data teams work might not only be surprising to them but also critical for them to have the context to collaborate well.

Table 5.5 illustrates some of the 'data team basics' that may help you paint a clear picture of the data supply chain for your business stakeholders.

Exhibit 5.1 Data-related spending breaks down into four areas (Source: McKinsey, https://www.mckinsey.com/capabilities/mckinsey-digital/our-insights/reducing-data-costs-without-jeopardizing-growth)

	1. Data sourcing	2. Data architecture	3. Data governance	4. Data consumption
Description	Cost associated with procuring data from customers,[1] 3rd-party vendors, etc	Cost associated with data infrastructure (procuring software, hardware) and data engineering (building and maintaining infrastructure)	Cost of data-quality monitoring, remediation, and maintaining data-governance artifacts (eg, data dictionary, data lineage)	Cost associated with data analysis and report generation (including spending on data access and cleanup)
Components	3rd-party data	Labor, infrastructure, and software	Labor, software	Labor, software
Typical owner of spend	Head of business unit	CIO	Chief data officer	Head of function or business unit
Typical spend, % of IT spend	5–25[2]	8–15	2.5–7.5	5–10
Example for a midsize financial institution,[3] $ million	70–100	90–120	20–50	60–90

[1]Excludes internal data-capture processes.
[2]Industries that don't directly touch consumers (eg, consumer packaged goods) spend a higher share (>20%) on data sourcing.
[3]For midsize organisations with revenues of $5 billion to $10 billion and operating expenses of $4 billion to $6 billion. Absolute values vary by industry and size of the organisation; eg, absolute spend is, on average, higher for the telecommunications industry.

Table 5.5 Data team basics for business stakeholders

The value and skills of data specialties Data specialties bring added depth to analytic and insights conversations where business context and judgement matter.	**Business analysts** use data to make business decisions, interpret and relay data information to stakeholders, and help solve problems. Specific skills and knowledge include mathematical process skills, statistical analysis methods, database management, regression modelling, reports production and presentation. **Data scientists** focus on the foundations of data science, including data systems, algorithms, data analytics and significant data infrastructure. Specific skills and knowledge include mechanics of data science tactics and methodologies, technological theory, engineering, statistics, algorithms and data structures. **Data engineers** understand the methods and technologies required to manage big data warehouses and engineering and problem-solving skills based in big data solutions. Specific skills and knowledge include optimising data collection and storage, processing and analysing data, reporting and visualising statistics and patterns, building and testing models to manage large quantities of data and programming languages. **Database management and architecture engineers** design and manage big data systems and curate and process data, and help others with organisation access and interpreting information. Specific skills and knowledge include database design and management, data warehousing, business intelligence, big data systems engineering, advanced architecture design and business analysis. **Data mining and statistical analysts** combine mathematical and statistical study with in-depth computational and data analytic training. Specific skills and knowledge include mathematical and applied statistics, data management, analysis and model-building with large datasets and databases, statistical computing and statistical learning. **Machine learning analysts** use advanced computational algorithms to expand cognitive functions and advance automation, and use algorithms and models to solve complex problems. Specific skills and knowledge include applied statistics, data mining, information visualisation, natural language processing, data and network security and pattern recognition.
Data's role in organisations If you hate maths, terms such as 'data', 'quantitative analysis' or 'pivot table' might sound intimidating.	**First and foremost, data exists to improve quality of life.** Used constructively, data increases awareness, measures/monitors progress and enables action – all of which should result in a better experience for the end user and the organisation as an interdependent ecosystem. **Data equals knowledge.** Good data provides indisputable evidence, while anecdotal evidence, assumptions or abstract observation might lead to wasted resources due to acting based on an incorrect conclusion.

(Continued)

167

Table 5.5 (Continued)

Data does not have to be complicated; it must be helpful to providers and consumers alike.	**Data keeps molehills from becoming mountains.** Monitoring the health of systems, processes and people increases the ability to respond to challenges before they become full-blown crises. Effective quality monitoring allows the organisation to be proactive rather than reactive and will support the organisation to maintain best practices over time.
	Data allows organisations to measure the effectiveness of a given strategy. When strategies are implemented to overcome a challenge, collecting data will enable leaders to determine how well the solution performs and how the decided approach needs to adjust over time.
	Funding is increasingly outcome and data-driven. With the shift from funding based on services provided to funding based on results achieved, it is increasingly important for organisations to implement evidence-based practice and develop systems to collect and analyse data.
Approaching a data strategy with intention An intentional data strategy doesn't happen as a discrete project. It requires the trinity of people, processes and platforms to align with the data strategy's goals.	Timing is everything for a data strategy to be successful. The growth phase of an organisation begins as it establishes a foothold in a market and develops its product or service offering.
	As the business becomes established, competition intensifies, leading to maturity as the organisation strives to achieve higher efficiency levels through cost minimisation and quality control. During this phase, functions are more independent, trying on their own data to make decisions.
	As markets become saturated, further growth becomes more complex, requiring new strategies (i.e. acquisition, diversification into emerging markets or development of new products). This increased complexity requires functions to increase collaboration across silos, share perspectives and make more consensus-based decisions. This new way of operating requires data and its underlying architecture to be restructured.
	An intentional data strategy enables multiple people on different teams or at different companies to work together with the same data on the same analysis to share and exchange context to build richer reports and faster business decisions – on purpose.
	Moving from ad hoc reports to an intentional strategy means the organisation engages in structural realignment to support the new strategy better. Management information systems are also being reviewed and developed to meet these new requirements. There is an increased emphasis on accountability, efficiency and cost savings.

Learning data jargon

Each discipline has a unique language. Outside those domains, vocabulary can sound like jargon. Everyone is guilty of using (and abusing) terms from the data sector. What phrases do you use frequently that might be confusing to business stakeholders? What phrases might they use that could be new to you? And what phrases are common to data teams *and* the business but also subject to misinterpretation? See Table 5.6 for some examples to discuss.

Table 5.6 Sample jargon by group

Business stakeholder terms	Shared jargon	Data team terms
Patterns	Best practices	Big Data
Net-net	Deliverables	Governance
Robust	Core competency	Fabric
Productise/Operationalise	Shared values	Artificial Intelligence (AI)
Space	Leverage	Machine learning (ML)
Solution	Scalable	Deep learning
Best of breed	Buy-in	Analytics (of any kind)
Storytelling	Development	Integrity
Information value	Infonomics	Cost (or data)

There's also jargon seemingly designed to confuse – buzzwords that both business and data teams use, but with drastically different meanings. Examples are provided in Table 5.7.

Neither interpretation is wrong. To ensure common awareness, err on the side of over-communication. Pause to ensure understanding.

Table 5.7 How jargon is interpreted by business and data teams

Buzzword	How data teams think	How business stakeholders think
Sustainable	Data teams think about *sustainable operations management,* which concerns procedures, processes, practices and systems that initiate, create and deliver profit while considering preserving or even improving the natural and/or social environment.[72]	A solution, service or product that is 'environmentally sound' or leads to 'reliable revenue'.
Mission-critical	Something that impacts the 'critical path' to the business roadmap.	Something 'critical to core operations' or the technical roadmap.
Performance management	The goal is to 'evaluate process, procedures, practices or systems outcomes'.	Where the goal is to 'evaluate employee performance'.
Programmatically	Where the goal is to 'systematise or automate' processes, procedures, practices or systems.	The goal is to 'align with other (business) programmes'.
Technical assistance	Where the goal is to provide for 'infrastructure support'.	Where the goal is to provide 'assistance with technology'.

---

72 Poul Houman Andersen, 2019. 'Sustainable Operations Management (SOM) Strategy and Management: An Introduction to Part I', in Luitzen de Boer and Poul Houman Andersen (eds), Operations Management and Sustainability (Springer 2019).

Keeping everyone motivated throughout the project

Think back to the last big project you took on. There were likely some peaks and valleys in excitement, motivation and energy. Similarly, your stakeholders will be in different moods at different points in the project. As a good data lead, you'll want to make sure you are aware of these ups and downs, empathise genuinely and support your stakeholders through the lulls.

Figure 5.2 shows some common trends, including the most likely points of disengagement specific to each phase. Still, it would help if you emphasised a consistent and continuing recognition strategy overall. When you work with various stakeholders, make sure to call out positive behaviours throughout the year, not just at the project's evaluation. Continue to provide value to your data team by connecting its work to the organisation's overall success and impact on the project's charter.

Integrate ethics throughout

Remembering the Data Oath we took in Chapter 1, this is a time in the project scoping process when we ask ourselves *can we* versus *should we*. We need to continue to ask that question throughout the delivery process, as a filter.

The impact of data insights can be harmful if it:

1. has a profound influence over a person's life (for example, banking); or
2. has even a shallow influence over masses of people (for example, social media).

Whether or not decisions directly impact another human being's life, every company needs to think in terms of the unconscious and *unconsidered* bias in their algorithms. The impacts are not just on the organisation's brand but on societal health.

A data breach has the potential to be much worse than an oil spill – harming people, weakening national security and disrupting business processes among global stakeholders. We cannot start to accept the known risks of data breaches as part of a solution's overall delivery costs as we do with the known risk of an oil spill.

Analytics change how decisions are made. *How will AI impact your proposed solution?* Is bias transparent? *What checks and mitigations do you have in place to determine the bias in your data?* Facial recognition is still evolving, but impacts are already felt as communities are policed differently. We know that credit, benefits and penalties differ for women and men. Before, technicians focused on the how, and leaders considered what should be done. Today, it's *everyone's* responsibility to ask: should we?

Every company has customer data. The question is, what can they do with it? When does knowing your customer go too far?

Figure 5.2 Team motivations: what are they thinking?

What are they thinking?	Prepare	Kick-off	Discovery	Building	Delivery and implementation	Evaluation and acknowledgement
Data team	This is needed; the scope will become clearer once we learn more about the real issues.	So much needs to be done; stakeholders will need to engage to pull this off.	No one starts from 0; great work has been done that we can build upon.	This is harder and more expensive than we expected; the scope must be adjusted.	It needs to feel like our regular job, not an additional burden. Identify current organisational pressure points to leverage.	Done! This project will make an impact!
Internal stakeholder	This is needed, but I'm unsure what I'm signing up for!	This will be 15 per cent of my annual goal.	It's heartening to see that we are starting with some momentum and can convert earlier efforts into learning.	This is bigger than anticipated and will require more effort; re-evaluate against other priorities.	Help identify existing processes, practices and systems to leverage for integration.	This project will make a measurable impact on my unit!
External partner	Seems exciting. I hope this meets our expectations and we can get some mileage from it!	Participation will be laid out in a contract or MOU.	We are excited to learn what we might apply to our own efforts.	With adjustments, our ability to leverage learning increases/decreases.	Help simplify the message and tell the story.	This project helped tell a story about what we are also going through and how we contribute to the solution!
Offshore resources	My part is clear; the scope will become clearer once we learn more about the real issues.	Contract confirmed.	Additional risks are outlined, evaluated and prioritised.	Adjusted scope impact timelines and budget.	Complete deliverables.	Evaluate satisfaction and performance against the initial contract.

High

Low

Motivation levels

Often companies think addressing ethical issues will be unfavourable for the business, or somehow 'break' what they do. But that is rarely the case. Embracing broader initiatives like sustainability, ethics, inclusion often yield exciting opportunities for iterative improvements.

It is difficult to evaluate risk and define required measures for use cases for which solutions have not yet been designed, let alone built. Nevertheless, the ethical risk is a crucial criterion in prioritising use cases and choosing an approach.

To that end, IG&H have created an Ethical Risk QuickScan to be used during the early stage of use case selection and requirements gathering. It gives a feel for the level and areas of risk involved early on in a project and, therefore, in which domains extra attention is necessary.

IG&H specialises in complex end-to-end digital transformation; the framework is reprinted in Figure 5.3 with their permission.[73]

The use case represents a sample of the thinking involved when determining checklist questions: imagine a taxi service that digitally monitors many aspects of the rides that are assigned and carried out by their taxi drivers. They want to develop a model that will score driver performance based on this data and automatically adjust driver paycheques accordingly. Table 5.8 helps us think through the framework.

Table 5.8 Applied practice: ethics fast-checks

Point your compass in the right direction as early as possible. Assess your data project or solution with this ethical framework as a team.	
Vulnerable people impacted?	The people impacted typically have below average financial means, and some may live paycheque to paycheque.
Number of people impacted large?	This internal application does not impact many people (only the company's taxi driver employees).
'Matters of life' affected?	The model's decisions will affect matters of life since they will impact financial well-being and job security.
Influencing personal behaviour?	The model's decisions may influence the behaviour of the taxi drivers in the sense that it might stimulate them to work longer hours or to accept fares to or from locations they are not comfortable with.
No, or slow, feedback loop?	The effect of the model's decisions will be visible quickly (every week or month), so there will be a fast feedback loop.
No human in the loop?	The use case calls for autonomous performance scoring and paycheque adjustment, meaning no human is in the loop.
Bias in the data?	Performance is typically subjective, so the data the model will be trained on may be biased.
Personal data used?	Finally, the data will involve fine-grained personal location data, considered personal data.

---

[73] Ethical Data Analytics, IG&H. IG&H distilled the Ethics QuickScan guidelines from a meta study done on 36 prominent AI principles documents.

Figure 5.3 shows how the QuickScan would be filled in based on this particular use case. For each of the areas that have outermost dots filled in, actions to mitigate potential issues must be considered or taken. Additionally, it is recommended to pay extra attention to areas that have the 'Medium Risk' dot filled in. Table 5.9 provides a sample of possible mitigations for the biggest trouble spots in this case.

Figure 5.3 IG&H's QuickScan applied in a use case

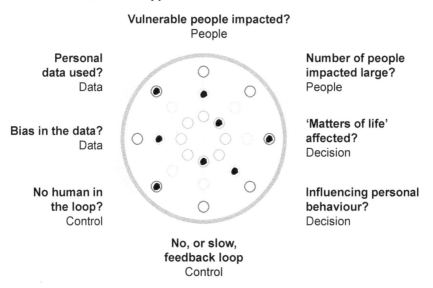

Table 5.9 Sample ethical QuickScan, applied

No human in the loop	• Analyse model sensitivity/robustness
	• Build model monitoring in implementation
Personal data used	• Ensure data is encrypted
	• Perform data privacy impact assessment
	• Strict anonymisation of data
Matters of life affected	• Periodically survey drivers about the impact of the algorithm
	• Monitor driver satisfaction and reward distribution

Filtering a data project solution through an ethical filter focuses attention on measures and mitigations that can be taken before a project is even under way. Once a project is in full swing, and as the actual solution and data requirements become increasingly apparent, the team can take its evaluation to the next level with the Ethical Risk

Evaluation Framework. This further examines at-risk dimensions in terms of the commonly accepted ethical guidelines of safety, fairness, transparency and privacy.

Phase 0: Preparing for kick-off checklist

Before moving to the kick-off phase, double check that you've completed the following:

1. Recall early discovery efforts from Phase Two: Determine Resources, specifically the answers to discovery questions for an executive sponsor. Integrate that information into the project planning process and communications.

2. The project scope has been clearly defined, confirmed by multiple stakeholders and captured in a scope template. It is up to you to ensure the scope is finalised and communicated to the team. Clear success measures have been defined.

3. Clear roles and responsibilities exist for internal stakeholders and sponsors, external partners, data team members and offshore resources. These roles and responsibilities should be a confirmed percentage of participants' annual performance goals to ensure accountability.

4. With your internal data team, you've confirmed a general timeline for completion, including work breakdowns of all deliverables. Tasks and deliverables should be assigned to working groups, subject matter experts, stakeholders and so on.

5. For each role on your internal data team and for internal stakeholders, you have confirmed that the assigned individual is prepared for the project kick-off and ready to supply the necessary management or feedback time (and that nothing about workload or availability has changed since your initial discussions).

6. For each role in the stakeholder group, you've confirmed that the assigned individual is prepared for the kick-off and ready to supply the needed time (and that nothing about workload or availability has changed since your initial discussions).

7. The executive sponsor and sponsor group have shared the project scope and intended outcomes.

8. Consider how many rounds of feedback to build into the milestone calendar. Consider what to ask and how the information should be collected, synthesised and shared.

9. Have you done an ethics 'fast check' review against the initial scope?

10. Send thank-yous to any referring individuals who have helped identify roles for the team, and let them know that you'll be kicking off the project soon.

11. For projects leveraging external partners, sign and return any memorandums of understanding, nondisclosure agreements, liability waivers or other paperwork related to the project.

12. Background materials and general organisational materials have been supplied to the data and stakeholder teams to help jump-start their understanding of the charter (and keep them engaged if there's a pause between involving them and kicking off).

1. KICK-OFF

With the preparation and planning phases complete, you're ready to move to the formal start of the project – the kick-off. Kick-off isn't typically a phase, but there are several activities to consider for a successful kick-off – activities which many people skip over. Kick-off focuses on preparing for the first meeting, bringing together the internal data team and the working and sponsor groups.

The data team lead meets with both the data team (including offshore, as appropriate) and working group to:

- confirm the data team roles and review responsibilities; confirm the percentage of time (and language) for inclusion in their annual goals;
- establish data team expectations;
- review project outcomes and expectations;
- agree on the final language the group members will add to their performance commitments.

Because this is a data project, the data team lead will manage the final product/service/ solution unless or until a handoff should be made. The working group will need to remain engaged less frequently for ongoing requirements and feedback. They might also require some level of technical assistance as updates get implemented. The point? *Keep a service mindset throughout the project – data projects are part of a sustained service.* Use Table 5.10 to review key data elements as a service. Discuss and continue to remind the team of the sustainability of the service throughout the project's lifecycle.

Table 5.10 Important implementation questions

Question	Example
What communications are needed to maintain stakeholder awareness and desire?	Stakeholders have requested a self-service reporting layer. The data team will collaborate on requirements and critical milestones. Reports using shared metrics and standard templates will be prioritised first. The data lead must ensure consensus or awareness of the plan as the data team moves through execution.
What systems or data might the staff and/ or stakeholders need to implement the solution successfully?	Example: If seeking to launch a new reporting layer, access to an existing web platform and support of any web developers, back-end database developers and data scientists would be critical to implementing a new portal.

(Continued)

Table 5.10 (Continued)

Question	Example
What should the review cadence be to keep the solution updated and relevant to the business?	Example: After the solution (reporting portal, data strategy, etc.), there should be a new governance committee consisting of the data lead and relevant business stakeholders. Ongoing requirements, feedback and access/usage policies should be determined and communicated.
	Take note: Engagement in ongoing, structured discussions and reviews of data project deliverables is the linchpin of data maturity in any group or organisation. Disciplined review of strategy leads to updates in strategy and increased commitment. Disciplined review of reporting portals leads to ongoing business requirements, increased usage and relevancy and impact on the business. Yet, many data teams do not host governing activities.
What training will staff and/or stakeholders need in order to use or implement the solution?	For a new reporting layer, the data team lead should work with data staff members (data scientists and data engineers) to develop training for business stakeholders (and extended stakeholders). End users must understand how to navigate the reporting portal and the reports hosted there. They need to learn how to interpret the data for themselves. Training should include helpful guides and tutorials about interpreting data, developing a data story, asking 'good' questions, analysing basic concepts and handling follow-up questions. The data lead should also manage the ongoing maintenance schedule for report design, accepting report requirements and key maintenance tasks (such as report updates).
Who might need to be educated and informed about the new solution to ensure its success?	Example: A data project to embrace a single data management architecture for the division should be shared with the sponsor group and executive sponsor to ensure a single source informs their business decisions of truth. The data lead should deliver this strategy proposal.

Data team kick-off

At this stage, the most important thing for you is to ensure that resources from the internal data team (on which you sit or lead) agree on the project scope and goals and can accommodate this project. In your internal kick-off meeting, you'll discuss the following topics:

- Review the project scope and initial need for that type of support (remember key idea 1 for making data management work: know and define needs clearly):
- Is the need still valid?

- Has anything changed related to the need that would impact the goals or stated outcome of the project? For example, a change in organisational priorities (like a significant shift in the product roadmap) means day-to-day engineering and marketing priorities might need radical adjustment. This could impact data needs requirements, and potentially the data project's key deliverables.

- Review the 'Roles and Responsibilities' template (See Table 3.8 and remember key idea 2 of doing data management work: get the right resources for the right job).

- Is each staff member with an assigned role prepared to meet the stated duties?

- Confirm that each team member:

  - Has the time to manage their part of the project effectively.

  - Has the required technical expertise to effectively support the data project manager in achieving the project scope.

  - Understands the impact of the project enough to stay motivated to complete the project work effectively.

  - Has not had a change in the volume of other work or priorities on their plate since they agreed to participate in the project.

- Review the working group and sponsor group structure with the executive sponsor for input.

  - Does the executive sponsor still have the authority to sanction this work?

  - Are they willing and able to help cultivate and maintain desire among their peers (the sponsor group)? Will they be active and visible?

  - Has the desire for the project been evaluated for both groups, and risks identified?

- Determine the feedback process for the data project team (remember key idea 5: learning goes both ways).

- How will you share timely and constructive feedback with the team?

- Review expected response times for replying to emails and submitting feedback. As a rule, a response no later than 24–48 hours after receiving an email is considered reasonable.

- Note team members' paid time off (PTO) and ensure you have proper coverage so work does not come to a standstill.

All these things might seem like common sense. However, it is *incredible* how often many fundamental concepts get overlooked, resulting in lost productivity, frustration with team members, increased cost and poor performance. When operating quickly, just trying to get things done, we can use shorthand – thinking that we have the same understanding of a group norm (such as response time or the importance of documentation) – and find ourselves stuck. We've all seen what happens when we don't do these things:

- A Vice President of Operations is completing a merger. They assign a project manager to lead a systems integration when the manager has confirmed three weeks off during a critical delivery phase. Awareness of this arose one week before the manager was to leave town.

- A Chief Data Officer adopts the charter of driving the data strategy for the organisation yet provides no vision or strategy guidance to support the work of the internal data team. Similarly, they attend no internal data systems discussions with sponsor-peers to help escalate issues or cultivate a desire for data transformation, preferring to focus on external sales calls and marketing activities.

- A senior data architect stops responding to email requests for his expertise, citing other priorities. He does not nominate a stand-in for the data discussions, and decisions to decommission data systems under his control are confirmed without his input.

- A general lack of documentation on a project forces the team to rely on verbal updates from meeting to meeting. The latest status resides with the last discussion on the topic – making feedback or insight hard to provide.

Poor communication, poor executive sponsorship, changing priorities and lack of procedure are behaviours many of us have observed or directly experienced. They can be mitigated by being thoughtful and intentional with how and why people are brought together.

Stakeholder kick-off

The project lead gathers the business stakeholders and sponsors, having confirmed the data team resources. The most important thing here is to ensure that working group members (across partnering groups, divisions or the whole organisation) agree on the project scope and goals and can accommodate this project *with a confirmed percentage of their annual goals*. In your stakeholder kick-off meeting, you'll discuss the following topics:

- Review the project scope and initial need that led to the scope of work.
- What do you already know about the unique nature of the project?
- What else do you need to know to be well set up for the kick-off?
- Review the 'roles and responsibilities template' (see Table 3.8).
- Does each stakeholder have a clearly defined role, with concrete responsibilities and an estimated time commitment?
- Is each stakeholder prepared to meet the stated duties? Confirm that each stakeholder:
  - Has the time to manage the data project effectively.
  - Has the required technical expertise to support the team effectively.
  - Understands the project's impact on the organisation and stays motivated to connect the project to broader data strategies.
  - Has not had a change in the volume of other work or priorities on their plate since they agreed to participate in the project – competing commitments are equally important to monitor with business stakeholders (and external partners) since they are not part of your direct reporting chain.
- Determine the feedback process for the data project team (remember key idea 5: learning goes both ways).
- How will you share timely and constructive feedback with the stakeholder group?

- Review expected response times for replying to emails and submitting feedback. For executive feedback, this can vary. Work with administrative assistants to determine the best windows for tight turnaround topics.

- Note sponsors' and stakeholders' paid time off (PTO) and ensure escalations, decisions, policy changes and budget updates do not come to a standstill.

Again, although these activities might seem elementary, they are fundamental to success. Some leaders get away with playing fast and loose – securing resources, claiming sponsorship authority – right up until they can't. Relying on shorthand that worked in other cultures and assuming what worked in one organisation will work in another is a common misstep:

- A vice president of product agrees to become a sponsor. They loop in three additional vice presidents of product, creating a lack of clarity for escalations and accountability.

- A Chief Data Officer sponsors a transformation effort but is new to the organisation and cannot assist the project manager in identifying stakeholders. An unusually high volume of stakeholders results in unclear buy-in of the scope, ineffective escalations and little accountability.

- The initiative can cease if the executive sponsor leaves the organisation and the sponsor group is unclear. The work might continue but be driven out of another team because awareness and alignment were unclear.

Sponsors are essential during organisational change because their involvement sends a strong message to the rest of the organisation. If the sponsor actively and visibly supports the change, individuals in the organisation notice. If the sponsor announces the change and disappears, individuals in the organisation notice that, too. The effectiveness of the sponsor is one of the strongest predictors of project success or failure.

An absent sponsor leads to consequences. Resistance to change increases among impacted employees. The team encounters more obstacles and cannot resolve critical scheduling, resources and scope conflicts. Changes are often delayed. When a sponsor has not built the necessary coalition of support at their peer level, changes fail to move past the pilot stage and lose momentum across the organisation. These consequences erode the value of a change, including return on investment (ROI). Given large data projects' importance in an organisation, the challenge of a missing sponsor must be addressed.

Reality check: Considering what the organisational culture expects and will tolerate, sponsors and change are essential. In some cultures, relationships are more important than the value any structure or operationalisation can bring, and it doesn't matter what the data says. Here, the 'soft power' of demonstrating a strong pilot is used to 'wake people up'. The hope is that people see how things need to change and feel eager to achieve this new vision themselves. Hope is not a strategy; these approaches are often short-lived and fizzle out.

Understanding what level of change is achievable and how quickly you can instigate change is critical to your success. Observe when the ground for change is fertile. Whether you are a CDO leader or a manager of data projects, temper the guidance, checklists and worksheets in this guide to cultivate the ground around you. From there, confirm your sphere of influence for change. Always be thinking of the next *reasonable* step forward. Change might start from the top down, from bottom up or somewhere in the middle.

Joint kick-off

At this point, each group has invested time separately, and you have likely had regular communication with your executive sponsor. Now you're ready for the formal 'Go!' The joint kick-off meeting serves several purposes:

- generating excitement for the work and its potential impact;
- confirming the data project purpose and process – getting alignment on scope, roles and timeline;
- setting up an effective working relationship, including communication expectations;
- getting to work – beginning to discuss the discovery phase.

Ideally, this meeting should be in person and onsite at the office. Kick-offs are an ideal opportunity to develop rapport and give a chance to connect to people who don't normally get together. The stakeholders will learn about your project scope and its impact on the organisation, will become more familiar with your data team and will feel more comfortable as part of the group. If an in-person meeting is not possible, think of ways to make the virtual experience engaging by adding polls, breakouts and opportunities for back-and-forth conversation throughout the experience.

A weak kick-off meeting could lead to:

- Unrealistic expectations about what a successful project looks like
- Confusion about the desired outcomes or deliverables
- Misunderstanding about the expected work process, communications and turnaround time
- Lack of commitment or buy-in from working group or sponsor group

A strong kick-off meeting could lead to:

- The right work completed at the right time
- A firm understanding of how to achieve success
- Trust and clear communication
- Greater impact on the community
- Greater sponsor and working group support and satisfaction

The joint kick-off meeting is the official start of the project work. The main objectives for this meeting should include:

- Creating a shared vision of success – including both shared goals and any separate goals for the stakeholders or data team.

- Ensuring that everyone understands the roles and responsibilities of each team member.

- Communicating the project's alignment with the organisation's overall goals and the expected long-term impacts – feed the stakeholders' motivations.

- Aligning on communications expectations – how will you interact? What's the best way to give and receive feedback for each team?

- Discussing the direction of the research coming up in the next phase, Discovery. Identify the key questions or potential interviewees. This could get you off of to a faster start.

Exhibit 5.2 provides a sample kick-off agenda for your use. This sample agenda is very detailed, and with good reason. Kick-off is a critical phase: many projects stall or go off course because they skipped a thorough discussion of goals and expectations. Feel free to work with your data team and stakeholder groups to adapt this agenda, and remember, this is your moment to set the stage. Use it.

Exhibit 5.2 Sample kick-off agenda

Topic	Cover
Team introductions (30 mins)	• Review and approve meeting agenda • Introduce the data team members – name, role, responsibilities on project, hopes for the data project • Introduce the stakeholder (working group) members – name, role, what they plan to contribute to deliverables, motivations for wanting to work on the data project, hopes for the engagement • Go around the meeting room and have each person state, 'At the end of this project, I will feel most successful if _____'
Project overview (45 mins)	Data lead leads the discussion: • Data team's overall charter, and how the charter of this data project aligns with organisational goals • Major challenges this data project addresses • A look ahead to ongoing sustainment processes, procedures or platform needs once data project is complete Hint: Now is the chance to show the stakeholder group what the data team provides for the organisation and how their contribution will make it better. Stakeholders can help by (reiterate their roles and responsibilities): • Asking clarifying questions • Drawing in their experience to ensure the challenge and potential solution are feasible, given constraints of the project (time, budget, resources, etc.) • Proposing alternative approaches or ways of thinking to meet the demands of the project

(Continued)

181

Exhibit 5.2 (Continued)

Topic	Cover
	• Maintaining reasonable expectations of the process, timing and impact of the project – providing feedback to refine expectations when needed
	• Co-creation of content or co-development of technology, as appropriate, based on resource needs
Project design (30 mins)	**Step 1: Share initial roadmap** Questions to address:

Step 1: Share initial roadmap

Questions to address:

• When should the project start and end? Are there specific deadlines that the working group is driving towards, and how is that built into the plan?

• Are there any external events that impact the roadmap and delivery of this project or are likely to impact stakeholders' ability to fully participate?

• What is the time commitment of working and sponsor groups? Is this estimate reasonable for everyone?

Step 2: Review working group structure

• What does the organisation chart look like?

• Who else in the organisation should be involved in this process?

Hint: Highlight the high level of commitment by showing the involvement of senior leadership.

Step 3: Review the required skill sets for this project

• What skill sets are you expecting the working group to have? Why were these people selected to be part of these groups?

• Compare your needed skills to available skills in real time. What is missing? Have any skill needs been overlooked?

Step 4: Review the process and implementation plan

• Review the work breakdown (list of potential deliverables, and their descriptions) by phase of work and desired outcomes.

• How quickly do you expect to hit each milestone?

• Review key risks/challenges and initial mitigations

• How will the final deliverable be implemented and sustained? What assistance will be needed from the working group and from the sponsor group?

• Are there specific resources, expenses or systems that will need to be modified or secured to use the deliverable?

Hint: For more tips on the requirements of successful implementation, see Table 5.10.

(Continued)

Exhibit 5.2 (Continued)

Topic	Cover
	Step 5: Set communications expectations and review logistics
	• Feedback: what do good/bad examples of behaviourally specific and actionable feedback look like? How will feedback be shared?
	• How quickly do you expect to respond to stakeholder inquiries? What is the preferred way to maintain status (newsletter, ticketing systems, email, all of these, others?). Are these reasonable to maintain for the project lead?
	• Will stakeholders need access to any particular systems, software or hardware during the project? What are the expectations of stakeholders in terms of their role and responsiveness?
	• What will be the course of action if stakeholders are unresponsive or if the project lead is unresponsive?
Next steps (15 mins)	Project manager to lead discussion • What are the key questions that need to be answered during Discovery? • Can you brainstorm, as a group, about which stakeholders or SMEs might be best placed to help answer those questions? • What are immediate next steps? • When is the next scheduled meeting for stakeholders, and the next for sponsors? • Are there any open questions that need to be answered before proceeding? • Close meeting by thanking stakeholders and tying the project back to the overall impact.

Reality check: It's possible this agenda will be too detailed for some organisations, and they might stop after learning the roadmap. Some cultures pay significant attention to how slides look, how personal the meetings feel when people gather; others might want even more information in the kick-off, and an even more operationalised agenda clearly labelling the decision-making style for each activity.

Always note what norms your organisational culture requires and tolerates for information sharing and gathering as groups.

Phase 1: Kick-off checklist

Before moving to the discovery phase, double-check that you've successfully completed the following:

1. The project lead has held an internal kick-off to confirm individual roles, responsibilities and availability from the data team to complete the project.

2. The project lead has held an external kick-off with the working group and sponsors to confirm roles, responsibilities and availability to uphold the commitment to the data project, including reviews and hands-on contribution.

3. The project lead has hosted a kick-off meeting (ideally onsite) for the full working and sponsor teams.

4. The concept of feedback has been discussed – its importance, what it looks like, where it will show up and how frequent it will be.

5. In addition to adhering to corporate data governance guidelines, the idea of using 'ethics fast-checks' has been discussed as a filter to use throughout the project. Concerns and risks will be added to the risk roster for discussion.

2. DISCOVERY

Discovery helps to identify better project scope and goals, resulting in more accurate budget and timeline estimates. Data helps facilitate better design decisions and ensures a higher return on investment. From a change management perspective, it is an inclusive process that naturally combats the narrow thinking that can happen when less diverse perspectives are involved in information gathering (that is, interviews and other research). It is designed to help the team learn as much as possible about the problem. Discovery enables a strong, collective understanding of the organisation's needs – both generally and in relation to the data project.

> *Remember:* The Discovery phase is more iterative than you think. It is a two-way dialogue between the project lead and the collaborating teams. As advice columnist Ann Landers reminds us, we can't 'accept the love of our dog as proof we are wonderful'.[74] We need the unvarnished truth from our stakeholders and partners about what is required and what needs to change.

This section will explore what to expect during this discovery period. We'll learn to ask the right questions of the right people and get the information we need to produce informed, well-designed solutions for stakeholders.

Providing contextual information

How do we set up a new team for success? We get clarity on who needs to be involved *and assess their level of support*. We provide information about the charter, known challenges, proposed mitigations and group operations – in other words, we orient them to the current context. We help them understand the current environment, constraints and opportunities in front of them through general background and project-specific items.

---

74 https://www.goodreads.com/quotes/135572-don-t-accept-your-dog-s-admiration-as-conclusive-evidence-that-you

Note: The first step in a discovery process is educating stakeholders about the context of the challenge you are attempting to manage or fix. Information about a critical issue area or operation can be highly engaging for the stakeholder team. Help feed their motivations and build momentum by being transparent with information and access to documentation.

General background

Table 5.11 gives examples of what you can provide to the stakeholder team to educate them on the team's charter, vision, goals and status.

Table 5.11 General background

Strategic plans, decision logic models, evaluations or diagnostics	Any existing strategic plan; logic models that informed decision-making, formal and informal evaluations that have been conducted (aka third-party audit or maturity diagnostics)
Background on problem, analysis	White papers or materials that describe the problem the organisation is seeking to address, approaches taken to address it and key trends
Organisation chart	Document that describes roles and responsibilities for project lead, stakeholders, sponsors, partners and others as appropriate
Meeting minutes	Description of what was discussed in recent meetings (open questions, decisions, escalations, new risks, etc.)
Dashboard, status reports	Reports or tools used to monitor programme performance
Research	Internal or external research (e.g. IDC, Gartner, McKinsey) on the issue area
Marketing artefacts	Any of the following documents relating to the productisation of the proposed solution to external customers: case studies, factsheets, summary of services, product feature comparisons and related web pages
Newsletters	Print or electronic newsletters
Peer organisations	List of 3–4 peer organisations that are aligned or competing for resources and funding with the data team
Key dates	Key dates that may drive a project timeline (such as external marketing events or leadership training/retreats)

Project-specific items

What background information is available related to the project? What documentation can you provide that will help the stakeholders understand the project at hand?

Think about the project's focus. What written documentation would provide a clear picture of this group/department/organisation, including previous evaluations of this issue? Table 5.12 offers an example of a data strategy project and the associated relevant documentation that might be helpful to the team.

Table 5.12 Example: developing a data strategy

Use cases	Develop a use case for each of your planned data projects. Use use cases to prioritise data projects and identify which ones to include in the data strategy. Strategic goals drive use cases. Examples of data use cases/data projects include:
	• Identifying new, smarter products or services
	• Automating processes to increase efficiency
	• Delivering a self-service experience (internal or external)
Data requirements	Data requirements address the questions: What are the highest impact business questions? What measures answer those questions? How will that data move through the supply chain (i.e. be acquired, cleansed, stored, analysed, shared and disposed of)? Requirements identify the common themes and issues that are related to the data itself.
	For example, a common theme across the data use cases might be data diversity. Specifically, how will internal or external data be combined, transformed and then structured to ensure the business question is answered with enough context?
Logical data model	A logical data model establishes the structure of data elements and their relationships. It is independent of the physical database that details how the data will be implemented. The logical data model serves as a blueprint for used data.
Logical architecture diagram	The logical architecture is defined as the organisation of the subsystems, software classes and layers that make up the complete logical system as it supports the data supply chain. It establishes the structure of data elements and the relationships among them. It is independent of the physical database that details how the data will be implemented. The logical architecture diagram serves as a blueprint for used data.
Data governance	This broad area encompasses data quality, ethics, privacy, ownership, access and security. As such, common issues are bound to be represented in the different use cases.
	For example, data quality might be an issue across the whole organisation. This means that before data priorities are achieved, there must be a continuous process to ensure all data is accurate, complete and current. Additionally, with cybersecurity threats rising, implementing a proven privileged access management (PAM) strategy is more important than ever.

(Continued)

Table 5.12 (Continued)

	The principle of least privilege (PoLP) is an information security concept that maintains that a user or entity should only have access to the specific data, resources and applications needed to complete a required task.
Technology	Determine software and hardware requirements across the data supply chain to support use cases.
	For example, consider two use cases that involve surveying customers at various stages of the customer journey (e.g. after they've contacted the customer service team or bought a product). Do you already have software to capture, store and integrate that data to gather insights? Or will you have to invest in new software?
Skills and capacity	Lack of data knowledge and skills is a big concern for many organisations. Use cases can address requirements around closing the digital and data skills gap, training stakeholders (or general employees or external customers), potentially outsourcing data collection and analysis where appropriate, or partnering with a data provider.
	If working with an external data provider, a use case requirement may be transferring knowledge from that external partner back into the company.

Observe. Observe. Observe.

Understand the current state clearly. Be behaviourally specific and descriptive (remove judgemental language). The first significant challenge is foundational: Where do we start? First, consider where you are today. Ask: What does good look like for the future? Ask how you expect future decision-making to move the organisation forward. For example, is the goal to drive digital transformation or provide a competitive advantage? Use Table 5.13 to help you assess current–future behaviours. Do not limit yourself to just assessing decision-making; consider other activities from the data supply chain.

Discovery interviews

When planning the Discovery process with your stakeholders, you want to strike the proper balance between depth and breadth in research by mapping out your potential targets for interview sources.

Step 1: Identify the key questions you need to answer in Discovery.

Step 2: Identify potential individuals internally and externally (where appropriate) who can provide answers to these questions. Can you get several answers from the same source?

Step 3: Research externally to gain a broader perspective of the problem you are trying to solve (that is, competitive comparisons, benchmark studies from consulting firms and so on).

Table 5.13 Current–future state analysis

	Current state of (decision-making)	Future (decision-making)
Example	• *Operational decisions, such as credit risk assessments and the next best offer to make, are manual, case-by-case and independent of decisions about how to service (or not) new and existing customers.* • *No consideration is given to the impact on resource availability to support these decisions.* • *This lack of transparency and connectedness hinders the organisation's ability to evaluate and effect strategic product mix and web commerce decisions.*	• *Decision-making is connected. More people are involved upstream and downstream to account for dependencies and collaboration.* • *Decision-making is continuous. Decisions become more automated, augmented and timely.* • *Decision-making is contextual. With increasing events and transactions, internal and external data sources are combined to create greater situational awareness.*
Your organisation		

Striking a balance between too much and too little discovery

Determining the right amount of discovery can be tricky. The project lead needs to get the right amount of information to deliver the right solution – especially regarding their organisation's unique situation or needs. At the same time, it's tempting for a project lead to let discovery go on too long to find a perfect or indisputable answer – or because they know less than the stakeholders about the problem area or key challenge. Table 5.14 illustrates too little or too much discovery.

Identifying the right discovery interview participants

A successful Discovery phase can make or break the project (see Exhibit 5.3). Without the proper information, the project lead risks creating misaligned recommendations or deliverables. Graduate students and those new to data management can fall into the trap of accepting discovery from an intermediary. For example, a project focused on employee onboarding should interview those recently hired as well as (but not instead of) the many stakeholders that deal with the newly hired. While business stakeholders who process newly hired employees may know of gaps in the experience, only the newly hired can shed light on the experience. This might sound like common sense, but substituting the end user's voice happens all the time in various contexts and can hamper accountability and progress.

Table 5.14 Excessive versus minimal discovery

Excessive

Minimal

	Too little	Just right	Too much
Risk	The team does not gather enough information to inform properly.		The team spends too much time investigating, energy stalls and the programme loses momentum.
Looks like	Less than six outreach participants, only internal or external sources, too many contacts from a single sponsor's network or organisational function, and too much focus on the current situation without organisational context or past performance.	A thoughtful outreach list, with internal and external sources representing a range of perspectives.	More than six interviews per stakeholder, more than one or two months' total duration, answering questions that are not core to project success, conducting interviews without a clearly defined objective and continuing to ask questions when the answers are always the same.

The project lead must be set up for success by having help in identifying the right individuals – both inside and outside the organisation – who will provide insights the team needs. Interviews should be conducted with people who provide:

- balanced and diverse perspectives;
- longitudinal and latitudinal organisational understanding (both over time and across functions);
- subject matter expertise;
- political influence, whether internal or external (are there people you need to talk to at this stage so that they buy into the project results?).

Using Worksheet 12, determine three primary questions you would like answered through discovery and then research who would be the most appropriate person to answer these questions. From there, build a list of interview participants to reach out to and base your brainstorming on the issues by asking them common questions.

Exhibit 5.3 Successful and unsuccessful discoveries

A successful discovery	An unsuccessful discovery
A project lead was tasked with developing a comprehensive data strategy for the organisation. The project lead interviewed various stakeholders, including third-party data providers and provisioners. Through these conversations, the lead identified a clear description of the extended stakeholders' needs. The data team was proactive in suggesting potential interviewees and made introductions quickly. This shortened the time spent on discovery and helped the data team strengthen those relationships through a well-planned process.	A project lead, assessing third-party data provisioners and possible collaborators for a central data warehouse, started their research according to plan. However, after a few exploratory interviews, the lead decided they had enough information to proceed. The team overlooked the three significant peer organisations, including an organisation that could have been a long-term advocate and partner. Because the suggestion was never made, the team missed a major opportunity. They also significantly overspent on the solution and were locked into a three-year contract.

Here are some standard guidelines that are relevant to any discovery:

1. There are typically four types of questions; if these are relevant to your project scope, make sure your discovery covers some of each:

   a. Simple questions are those that you know will lead to direct, relatively uncontroversial answers. Direct these questions to the day-to-day owners of current systems, processes and procedures.

   b. Strategic questions are relevant to the direction and impact of this project. Direct these questions to the senior staff, functional leads and sponsor interviewees.

   c. Audience questions explore the beliefs of the project's intended audience. Direct these questions to stakeholders and extended stakeholder representatives.

   d. Competitor questions regarding competitive positioning should be directed to external-facing staff (product managers, marketers, etc.), knowledgeable senior leads and 'competitors'. Sometimes, third-party partners are considered as competition when build-or-buy decisions are considered. For example, you might initially outsource a metadata solution. Use the opportunity to educate yourself on what requirements are important and then build the data solution internally.

2. Depending on the project type, information from 'competitive' organisations can be useful, including interviews. If the project lead does conduct interviews with competitors, they must remain honest in describing who they are and why they are calling. It is unethical to lie about identity and purpose. Conversations with partners also require political attunement. For example, you don't broadcast that you are getting smart on their technology, but most third parties can see which type of customer you will be based on your data usage and fluency with their features.

3. These questions can be asked face to face, over the phone, via email, in a survey – or using a mix of these methods – as appropriate for the project's particulars.

4. In some cases, interview participants will request confidentiality or anonymity. The project lead should use their best judgement in telling who said what. In most cases, hearing the correct information is more important than identifying the interviewee.

What discovery questions will be necessary for your data project? Use Table 5.15 to develop suggested questions for your working group. Use this table to address the key questions you mapped out from above.

Table 5.15 Questions for working group

Question type	Sample question
Simple questions are those that you know will lead to direct, relatively uncontroversial answers. Direct these questions to the day-to-day owners of current systems, processes and procedures.	What is the team's mission and vision? What is the budget size? Dedicated resources? What is the process for the current state (of whatever issue you are researching)? (i.e. how are things done today?)
Strategic questions are relevant to the direction and impact of this project. Direct these questions to the senior staff, functional leads and sponsor interviewees.	What are the top five goals for the team this year? How will this project advance those goals?
Audience questions explore the beliefs of the project's intended audience. Direct these questions to stakeholders and extended stakeholder representatives.	List the most critical stakeholder groups. How would these groups describe the organisation?
Competitor or partner questions regarding competitive positioning should be directed to external-facing staff (product managers, marketers, etc.), knowledgeable senior leads and 'competitors'.	Who are your top three competitors (external partners or data provisioners)? What distinguishes our internal solutions from theirs?

Other kinds of discovery

Depending on the complexity of your data project, interview methods alone might not be sufficient or appropriate. The main goal for Discovery is to provide the project lead and working group with the information, context and insights they need to develop the deliverable(s) that needs to be produced. Discovery might also include other kinds of listening:

- An environmental scan or review of external factors relevant to your organisation, challenge or project (such as recent legislation or media coverage).

- A review of printed or published materials.

- Training on key issues or insights into specific areas.

- Exploration of key hardware, software or organisational systems that will be affected by project results.

- Review of the materials, research or data from prior projects or studies.

- Understanding internal and external compliance guidelines.

These appear to be standard project management activities. Yet, when we face some of the biggest dilemmas of our time, they can have an incredible impact on individual customers and society.

Analysing your discovery results

After gathering the discovery information, the project lead will digest, synthesise and present initial findings to the working group and sponsors for feedback and additional insights. The project lead's recommendations are the first pass at analysis. Further analysis of the findings occurs when feedback has been integrated from the working team and sponsor groups, respectively.

The team uses the discovery results to determine the best direction for the upcoming build phase, making it essential to unearth assumptions being taken towards the next steps. In some cases, the feedback can help keep the data team from drawing the wrong conclusions from the findings. See Exhibit 5.4 for some common pitfalls in analysing data.

Exhibit 5.4 How objective are your decisions? (Source: https://en.wikipedia.org/wiki/List_of_cognitive_biases)

Even the best leaders and strategists make bad decisions due to personal bias. The project lead must receive but not react to recommendations from the stakeholders and sponsors. Additionally, they must challenge the working group and sponsor's thinking to mitigate chances of making irrational decisions.

Below are examples of common cognitive biases (ways we trick ourselves) that come up when making decisions at work. Which ones do you think you are susceptible to, given your personality, work style and experience? Ask your stakeholders which biases might trip them up.

Anchoring	The tendency to rely too heavily, or 'anchor', on one trait or piece of information when making decisions (usually the first piece of information that we acquire on that subject).
Availability heuristic	Estimating what is more likely by what is more available in memory, which is biased towards vivid, unusual or emotionally charged examples.
Bandwagon effect	The tendency to do (or believe) things because many other people do (or believe) the same.
Base rate neglect	The tendency to base judgements on specifics, ignoring general statistical information.
Belief bias	When someone's evaluation of the logical strength of an argument is biased by the believability of the conclusion.

(Continued)

Exhibit 5.4 (Continued)

Confirmation bias	The tendency to interpret information in a way that confirms one's preconceptions.
Expectation bias	The tendency to believe, certify and publish data that agree with one's expectations for the outcome of an experiment, and disbelieve, discard or downgrade the corresponding weightings for data that appear to conflict with those expectations.
Framing effect	Drawing differing conclusions from the same information, depending on how that information is presented.
Groupthink	The tendency for a group of people where the desire for harmony or conformity result in an irrational or dysfunctional decision-making outcome. Group members try to minimise conflict and reach a consensus decision without critical evaluation of alternative viewpoints by actively suppressing dissenting viewpoints, and by isolating themselves from outside influences.
Knowledge bias	Tendency of people to choose the option they know best rather than the best option.
Mere exposure effect	Tendency to express undue liking for things merely because of familiarity with them.
Ostrich effect	Ignoring an obvious (negative) situation. Avoidance.
Reactance	The urge to do the opposite of what someone wants you to do out of a need to resist a perceived attempt to constrain your freedom of choice.
Status quo bias	Tendency to like things to stay relatively the same.

Tips for success in analysing the discovery results:

- Make time to process the feedback and analysis from the working group and sponsors before providing recommendations.
- Look for areas that may be underexplored or not appropriately captured – redirect the team to take a closer look at these areas if needed.
- Push the team to ensure that they are:
  - making the information in the findings accessible (that is, concluding, not simply presenting facts);
  - reading between the lines;
  - following a standardised review process to understand the interview findings;
  - gaining different perspectives on the data – it may make sense for them to review individually before discussing as a group to ensure thoughtful, independent analysis.
- Give feedback to the team in the form of questions – not statements. Let them conclude on their own.
- Help them by asking for clarification on unclear areas.
- Never present analysis or recommendations on behalf of the team. Let them present the analysis in their own words.

Phase 2: Discovery checklist

Before moving to the build phase, double-check that you've completed the following:

1. The project lead, data team and working group have identified a limited number of targeted key questions to investigate through discovery outreach.

2. The data team and working group helped the project lead to find the right outreach targets for each key question and provided the necessary introductions.

3. The project lead conducted all the interviews with consistency.

4. The project lead completed all the discoveries, such as secondary research reviews. Working groups and subject matter experts helped as appropriate.

5. The project lead reviewed all findings and prepared a summary of insights, including an outline of the next steps and the build phase.

6. The project lead shared the discussed findings with the working team and sponsors, including the intended direction for draft processes, procedures or solutions (such as a data strategy or governance policy). What will the deliverable look like? What will likely be included, and how will the discovery inform that deliverable?

7. The working group and sponsors supplied clear and actional feedback on the findings and direction, and this feedback was shared and integrated into the final recommendations. Risks and open questions have been logged and will be tracked.

8. The group underwent an 'ethics fast-check' review to provide a forum for concerns and document mitigations.

9. The working group signed off on the recommendations for the sponsor group. The sponsor group signed off on the direction for the project's next stage.

3. BUILD

Once the Discovery results have been agreed upon and course corrections have been made, it's time to do the work. The Build phase is about creating, developing, improving, writing or producing the end process, procedure or solution. This phase looks different for every deliverable; there is no step-by-step process, but there are some best practices to consider to be successful. We'll look at two: storytelling and issue management.

Storytelling

The project lead and the data team must build their storytelling competencies to sell the benefits of data initiatives to the working group, sponsors and other business leaders. The story must be clear and concise, contain a headline and three supporting points, and list the overall benefits. Table 5.16 provides an effective storytelling outline, summarising the impact of reviewing the use of a third-party data warehouse.

Table 5.16 Storytelling outline

Three steps for success for effective storytelling with data	1. Objectives and KPIs	2. Process and decision impact	3. Benefits
	Identify objectives and key performance measures (KPIs)	Map data and analytics initiatives to objectives and identify processes, activities and decisions impacted	Identify financial and non-financial benefits and KPIs for success
Third-party data warehouse review, example	**Objectives** Reduce the cost of data storage and increase transparency of data usage **Measured by** • Cost of data stored • Percentage of data used vs stored	**Analytics initiative** Data supply chain visibility **Processes impacted** • Data storage and usage planning • Data inventory management	**Quantifiable impact** • Economies of scale from aggregating data in a single location • Reduced storage cost once used data is identified • Reduced waste (unused or duplicate data) **Nonquantifiable impact** • Improved data usage, quality and storage effectiveness

Once measures, processes and benefits have been identified, teams can fall prey to two common mistakes: (1) overdependence on progress metrics as opposed to business impact and (2) overreliance on numbers for a return-on-investment exercise to get projects funded. Table 5.17 illustrates how to consider these mistakes.

Issue management

The people side of change takes a significant amount of time. If open issues remain unresolved or there are objections from anyone, they must be vetted now. Suppose work starts before these concerns are mitigated or intentionally set aside. In that case, there might be frustration across all the stakeholders, members might begin to hold back their participation and it will be difficult to implement the work produced. Buy-in helps realise impact. There are three angles to consider: addressing member resistance, achieving consensus on how deliverables are revised and ensuring bi-directional feedback is balanced and actionable.

Table 5.17 Common storytelling mistakes

Over-dependence on progress metrics as opposed to business impact

Instead, translate clearly <u>defined</u> metrics into impact metrics:

<u>Faster</u> Time to Insight *leads to more* **Revenue**

<u>Increased</u> Data Quality *leads to decreased* **Cost**

<u>Reduced</u> Model Development Time *leads to decreased* **Risk**

Over-reliance on numbers for a net present value or return on investment exercise to get projects funded

Instead, dedicate more time to creating and telling a holistic value story than spiralling on the numbers. Data projects are more than cost-reduction strategies.

50%	25%	25%
Think about the value story	Get the numbers right	Tell the value story

Sometimes, group member concerns come through a variety of resistance behaviours. Group members might feel threatened and think they are being automated out of a job. When the electric vehicle emerged, people on the traditional manufacturing line thought they were working on 'old technology'. Ford leaders had to ensure that employees felt they were contributing to a bigger whole rather than simply working on 'old and less relevant' or 'newer and better' products.[75] See Table 5.18 for a description of resistance behaviours and how to work with them. Digital and data transformations are particularly challenging because the goals hit every aspect of the organisation: the customer experience, the workspace, operations and infrastructure and products and services.

Resistance shows up in many subtle and not-so-subtle ways, because to say that change is hard is an understatement – especially when emotions are running high. Change is also never the work of a single person. It takes a coalition, making executive sponsors essential. The sponsor group is not just there to beat the drum of support. Sponsors are instrumental in aligning organisation and transformation goals, helping to create consensus among diverse stakeholders, acknowledging positive progress to reinforce what they want to see, lining up the resources necessary for the project to succeed and serving as a constructive escalation channel for issues and concerns. Their role in the success of change initiatives often gets overlooked because we focus on what we can see: project execution and deliverables. Another reason why good sponsorship is hard to spot is that so much of it happens out of sight for most stakeholders. However, projects with a highly effective sponsor meet or exceed objectives more than twice as often as those with a very ineffective sponsor.[76]

Achieving consensus on how to revise or provide feedback on the deliverables depends on frequency and transparency. A schedule was determined as part of the scoping activity. Regular opportunities to be heard and reflected can inform risks and identify

---

75 'The Industrialist's Dilemma: Mark Fields, CEO of Ford', https://www.youtube.com/watch?v=1nOZ0zj7ILA

76 https://www.prosci.com/resources/articles/primary-sponsors-role-and-importance#:~:text=Prosci's%20research%20

Table 5.18 Different forms of resistance and strategies to mitigate (Source: Adapted with permission from Peter Block's *Flawless Consulting*)

Give me more detail	A person keeps asking for more detail and finer and finer bits of information. There is an insatiable appetite for every detail, and no matter how much information is provided, it is never enough.	When you start to become frustrated with the number of questions even though you can answer them, it is time to suspect that the questions stem from resistance rather than a simple request for information. *Name it*: *'You keep asking for different versions of the same document and are asking to be involved in every meeting. Is there an underlying concern you have not expressed?'* *Try*: questioning the question. What will the information answer? How will this help progress? What is your true underlying concern?
Flooding you with detail	A corollary to 'Give me more' is being given too much information. When asking how the problem started, someone will return to the company's founding rather than bracketing the issue and constructively framing the details to offer a simple story. The person keeps providing details, and it doesn't clarify things.	When you start to get bored or confused about what all the information means and how it connects, suspect resistance and not just a generous attempt to provide all the facts. *Name it*: *'You are giving me more detail than I need.'* *Try*: slowing them down. *'Talk to me like a 12-year-old. Please give me the executive summary of the issue. How would you simplify this issue?'*
Time	A person says they want to support your effort, but the timing is off. Next week or month will work better. They keep cancelling meetings and, when you do have a meeting, they allow many interruptions, or they multitask. They may be trying to communicate that they have an exciting job or have little time to spare, or they want you to think they are refusing you because of time constraints and not because your project makes them uncomfortable.	We all face time management issues, but time is most often used to avoid conflict, feedback or confrontation. *Name it*: *'You look like you have other things on your mind and little time for this project.'* *Try*: seeking alignment. Assess competing commitments and if those can be re-aligned. Determine whether this is an essential stakeholder, if there can be a stand-in or where the escalation path lies in ensuring participation.

(Continued)

Table 5.18 (Continued)

Impracticality	The person keeps reminding you that 'we're in the Real World', which is a veiled (not) accusation of being impractical, academic, theoretical – or, the ultimate sin, idealistic.	While there may be some truth to their perception of our idealism, as all projects must have at least some, the intensity of their mention of 'practical' should arouse suspicion of resistance.
		Name it: 'The proposals presented so far have been met with scepticism. What has been tried so far that has worked?'
		Try: empathy.
		Understand why they have this perception. Has something like this been tried before and failed? (Probably.) Good ideas never die. What is different about the current environment or timing? Timing and sponsor alignment don't always match up, but we can accomplish so much when they do!
I'm not surprised	Some people have a hard time giving or sharing credit. When one person thinks of a creative solution or identifies an intractable problem that others missed, there will be someone in the group who will not be surprised. Surprise might indicate that this person was not in the know or politically connected. Fear of surprise is linked to the desire to be in control.	You are instantly deflated, because we've been told by this behaviour that what we have proposed is not helpful, unique or necessary. It downplays our contribution.
		Name it: 'The ideas discussed don't seem new to you. Is that the case? How did they fail?'
		Try: agreeing.
		See this reaction for what it is: resistance, not a reflection of what was proposed.

(Continued)

Table 5.18 (Continued)

Attack	A person directly confronts us with negative emotion or feedback or unconstructive dialogue. Their skin flushes. They get loud, pound the desk, point fingers, and emphasise every. Point. They. Make.	When we feel awkward and fumble material we know inside and out, or feel like we have crossed some moral line, we might want to withdraw completely or respond in kind. *Name it*: 'You are really questioning a lot of what I do. You're getting loud, and I'm reading that as anger about something.' *Try*: not to take other people's reactions personally. How someone behaves reflects them (their skills, their pressures, etc.). See it as the resistance it is. Take note if you need to defend yourself (likely not), log the issue and promise to close out on this conversation later. Give the person and yourself some distance.
Confusion	Someone might come to us for help, needing clarity. After the issue becomes clear to you, you might reflect your understanding two or three times, but the person still does not understand the problem.	This might be resistance when you cannot find common ground through mutual understanding. *Name it*: 'You seem very confused about what we are discussing. Are you confused about the problem or just not sure what to do about it?' *Try*: breaking down the confusion. Understanding what other concerns they have about what they don't seem to understand or if they can think of adjacent issues to discuss. Perhaps looking at it from the angle of something you know they understand might help.

(Continued)

Table 5.18 (Continued)

Silence	A person was in contact with you but starts to cancel meetings. They don't reply to emails. We make overtures, but they don't engage. When they engage, they'll say, 'Everything is great!' 'Keep it up.' 'If I have issues, I'll speak up.' Blocking by withholding input is a kind of fighting stance. These behaviours communicate, 'I am holding tightly to my position; I won't even share my words.' Remaining silent can often backfire because their voice (and vote) never got to count for anything.	Silence never means consent. If the work is important to the organisation, it is not natural for people to have no opinion or thoughts on what is being decided. *Name it: 'You are very quiet. I don't know how to read the silence.'* *Try: being (respectfully) direct.* Ask yourself what concrete support this person provides and how they have shown enthusiasm or gotten personally involved in an action. Be behaviourally specific. If you can't come up with anything, wonder whether the silence is resistance.
Intellectualising	When a person shifts the dialogue from deciding the next steps to exploring theory after theory about why things are the way they are, this is resistance – of some kind. It could be their attempt to calm a 'fidgety brain', or they could be attempting to take the conversation off course. Regardless, spinning theories is a way of taking the pain out of the situation.	When you feel pulled into engaging in ceaseless wondering, when confronted with a difficult situation – especially when the intellectualising starts in a high-tension moment – try the following. *Name it: 'Each time we get close to deciding what to do, you go back to developing theories to understand what is happening.'* *Try: focusing.* Bring the discussion back to action, away from theories.
Moralising	A person presents an us/them dynamic with words such as 'that team/those people' (about people not in the room), or 'should' (suggesting a right/wrong or good/bad way of doing things), and 'they need to understand' (essentially, pedestal sitting). All these words point to the idea that this person is the one who 'gets it' and everyone else doesn't.	When you hear these words, you are about to go on a trip regarding how things ought to be. You might even get seduced into buying into this future picture. It might even make you feel more elevated, powerful and politically protected. *Name it: 'You have a clear vision of how things should be, but others aren't getting it for some reason. Why do you think that is happening?'* *Try: grounding in the present.* Resist the temptation with as much grace and persistence as possible. We are where we are, today.

(Continued)

Table 5.18 (Continued)

Agreement and compliance	A person agrees with everything you say and do. They want to know what is coming next. A lack of reservations about anything should be a warning sign that something is missing. If feedback isn't expressed to you directly, it will come out in some other less constructive way (gossip, backchannels, etc.).	When you feel agreement, enthusiasm and understanding of what you are facing you might feel lucky and not take it as resistance, even if a few reservations are expressed. *Name it*: *'You seem willing to do anything I suggest. I can't tell what your real feelings are.'* *Try*: proceed with caution, call attention to lack of resistance as a risk. Ensure that there has been a thorough discussion of the problems before a rush to solutions. Be cautious of people who act as if they are dependent on you and imply that whatever you do is fine.
Methodology	With elaborate data collection come questions about the methodology used. If it was a survey, there would be questions about the sample size and statistical significance. These questions are legitimate for about ten minutes. (It is assumed the method and results were vetted with a manager or colleague as appropriate.) Appropriate labelling and explanation of data will help establish credibility.	When interest and time dedicated to methodology increase past a reasonable time frame, suspect resistance. Repeated questions about methods or suggestions of alternate methods might delay the discussion of problems or actions. *Name it*: *'There are a lot of questions about my methods. Do you doubt the credibility of my results?'* *Try*: returning to the agenda. The purpose of the meeting was not to drill into methods but to review data points that might help the group understand the problem and decide what to do about it. Data is meant to help, not hinder, the development of perspective.

(Continued)

Table 5.18 (Continued)

Flight into health	Somewhere near the middle or end of the project, a stakeholder no longer has the problem the project is solving. As the delivery and implementation dates draw nearer and the stakeholders might have to act, things are suddenly improving. Variations on this theme include: (1) profits were down and people were feeling bad, and now they are up and people are feeling better, so there is no need to do this project; (2) a date is established for the project, and the stakeholder communicates that the problem has reduced in severity (but what really happened was that the problems are still there, the group just wants to ignore them). As deadlines loom, so do the behaviours that need to change – and people want to avoid that mindset shift.	Surface symptoms have underlying problems that require attention. For example, if there is long-standing conflict between groups and one of the leaders moves on, the conflict is still there. *Name it: 'As we approach implementation, the problem seems to have disappeared. What has changed (specifically)?'* *Try: stating the obvious, calling attention to the issue.* When surface problems appear to be clearing up, be cautious about grasping for improvement too early and smoothing over the problem's real focus.
Pressing for solutions	'I don't want to hear about problems; I want to hear about solutions.' The desire for solutions can prevent the project lead and the stakeholder from learning anything important about the nature of the problem. It also cultivates stakeholder dependency on the project manager to solve every problem.	When there is little tolerance to slow down and examine the problem, solutions will not be implemented effectively. *Name it: 'It's too early for solutions. I'm still trying to find out...'* *Try: grounding in reality.* Recognise that the rush to solutions can be a particularly seductive defence and form of resistance for those eager to solve problems. Slow down to speed up. Understand the complexity of the problem and then simplify.

opportunities. Synthesising feedback and sharing it appropriately with members helps to develop trust and ensures it becomes integrated into the process. There are then no 'black holes' into which issues disappear.

Ultimately, the project lead must synthesise and reflect diverse perspectives. They need to negotiate trade-offs. Not everyone will be pleased all the time. The lead must make choices and know when to be willing to be misunderstood. This might require the lead (and sometimes the executive sponsor) to invest in relationship repair and ensure everyone is still on board with the big picture. (See 'The Project Change Triangle and Assessment' in Chapter 4 to recall needed behaviours.)

Bi-directional feedback on how things are going must be balanced and actionable. Considering the good, the bad and the ugly is important – for the project lead, working group members and sponsors. The project lead is assessing everyone in the group for engagement and contribution, just as the group members are assessing the project lead's ability to deliver results. Most feedback is well-meaning, and members are open to being coached into giving the most constructive feedback possible. Hearing people into speech, when we know what we want to say but cannot find the words, is a skill we all need to develop. See Table 5.19 for examples of constructive and unconstructive feedback. At the same time, not all feedback is helpful or constructive. Sometimes it might be delivered too late in the process and thus thwart momentum. When creating the scope document with the working team, determining feedback rounds was cited as part of the planning process. If this activity was skipped, address it now. Ensure time is made for agreement and disagreement.

Table 5.19 Constructive feedback versus unconstructive feedback

Constructive feedback	Unconstructive feedback
• Comes in the form of questions, not solutions.	• Decides the direction for the stakeholders.
• Addresses things that can be changed, improved or adapted.	• Is not actionable.
• Is informed by multiple perspectives.	• Is not well organised – is passed to the team in pieces.
• Is developed in partnership with the data team members.	• Is given too late in the process, after many opportunities to provide.
• Is shared for consideration, and then actions are brainstormed in the group and synthesised by the project lead.	• Is inconsistent and/or radically changes the scope of the project.
	• Is not tied to the strategy of the project or the organisation (i.e. is random).

Data project 911

By this point, the project lead is well into the bulk of the project work. Even after mitigating all the resistance you can, challenges will present themselves. How can you handle difficult political and interpersonal situations without disrupting the project work? Table 5.20 identifies a few of the most common 'What now?' situations that present themselves, with possible responses.

Table 5.20 'What now?' solutions and what to do about them

A data or working team member goes silent and isn't responding to emails or phone calls.	If a stakeholder or working group member disappears and is unreachable after repeated attempts to contact, involve their manager to see if they have taken time off (e.g. vacation, child leave, illness, etc.). Make sure to track time off for all team members, ensuring stand-ins for continuous coverage to maintain progress.
The project lead cannot complete the project.	Determine how far along the project is, whether the responsibilities can be distributed across other data team members or if another project lead should step in. The latter solution might require a project reprioritisation, pausing other projects (or this one) or a new hire.
There is no time on the calendar that works for everyone.	Refer to previous kick-offs you have conducted. What were the communication preferences? Did everyone agree to meetings that would accommodate all time zones? Remind them of the commitment they made in joining the project and ask them to compromise a bit on finding a time that works for everyone.
	Try providing direct feedback if the challenge rests with just one individual. Let them know that when they aren't available, it signals to everyone else they aren't committed to the project's success.
	Depending on the size of the group and how aggressive the deadlines are, lock on a minimal meeting cadence. Start synchronous meetings with quorums. Make extra effort to conduct informal polls and provide more frequent updates (weekly, monthly, quarterly) and solicitations for feedback.
Members take too long (more than four days) to respond to emails.	Send a gentle reminder about the terms you agreed to at the kick-off meeting. Remind them how delays in delivering expected work will impact the timeline and eventually become a risk that requires mitigation. The project lead helps keep the group accountable to the norms that were agreed to.
The working group is not satisfied with the solution after multiple revisions.	Remember that the working and sponsor groups' role is to provide clear and actionable feedback to the project lead and data team. If there isn't a solution that meets expectations after multiple revisions, it could result in the project lead not receiving enough proper feedback. If timely and actionable feedback is given and things aren't headed in the right direction, speak candidly with the project lead and executive sponsor. Make sure there is space for a two-way dialogue about where there are issues so the project can still get back on course.

(Continued)

Table 5.20 (Continued)

Scope creep: the team is going out of scope and delivering things not part of the original agreement.	This might not be a bad thing! Maybe the data team found additional ways to support the effort. If the additions favour the solution, amend the scope document to reflect the new deliverable. Communicate the update to the working and sponsor groups.
	However, if the additions interfere with the original scope, refer back to the scope document and remind the data team of what is and is not meant to be accomplished. The work may still be valuable; create a separate scope document and add it to the project backlog to be prioritised later.

Phase 3: Build checklist

Before moving to the Implementation phase, double-check that you've done the following:

- Solicited (from stakeholders) and raised (your own) concerns, questions or objections regarding the direction of the process, procedure or solution as early as possible.

- Held at least one meeting where the first draft of the deliverable was shared.

- Informed key stakeholders about the project direction and status and asked for input on key decision points from sponsors and working groups.

- Scheduled and conducted a second meeting in which the project lead provided feedback to the data team and working group. Review of the deliverable and feedback on progress should be two separate sessions. This leaves time to digest, reflect and bring feedback to the team for conversation.

- Conducted the remaining rounds of revision and feedback as agreed to in the project scope.

- Ensured that the data team and stakeholders who are on point for deliverables have a clear plan and timeline for delivering the solution (in the next phase), including clear next steps on how to get there.

- The group underwent an 'ethics fast-check' review to provide a forum for concerns and document mitigations.

4. IMPLEMENT

Once the project lead presents the initial deliverable and clear and actionable feedback has been given by the working group and sponsor group, it's time for the final delivery of the process, procedure or solution. The project lead leverages contributions from the working team and SMEs and is responsible for the final deliverable. The lead ensures all necessary implementation steps and training are planned and completed. The working group and sponsor group still need to contribute feedback throughout this process. The lead and data team are available for clarification.

Tracking impact

The project lead delivers the deliverable in this phase. The final group meeting is the 'official' handoff of the deliverable, and it is important to bring the whole working group and sponsor group together for this presentation. The groups have seen and reviewed iterations of the deliverable before this point, so they have a good sense of what the project lead will present. If there are any surprises in this meeting, the deliverable is not delivered; you are still in the Build phase.

When making decisions, we are often not aware of the impact of previous decisions. Tracking and analysing the results of findings is one of the most challenging elements of decision-making. If we do track the results, we aren't always choosing the right measures. Regardless, tracking decisions presents the opportunity to re-examine good *and* bad decisions for optimisation and learnings. Use Table 5.21 to start to log and review project decisions.

Table 5.21 Applied practice: take a closer look at the decisions you've already made

	Decision	Metric	Learning or Best Practice
Identify decisions whose outcomes are not being tracked			
Identify decisions where the appropriate metrics are not being used to measure success			
Identify the benefits of re-examining good decisions to determine the best practices that should be repeated			

Analysing decisions, reviewing measures and taking stock of learnings keeps insights actionable and relevant.

The delivery meeting focuses on implementation. What is the plan for taking and using this deliverable? For example, how should the results of a data maturity audit influence the data strategy? How will implementing the first phase of the data strategy influence ongoing report requirements? How will you apply report usage patterns to data access privileges? This meeting also focuses on identifying continuous ownership or transferring ownership of the project to the group that will maintain the efforts.

If this work originated in the data team, it would likely stay there but could create an opportunity to align and partner with other data groups. If data resources are already consolidated, the ongoing maintenance of this work gets evaluated along with other sustainment efforts and feeds into the business's ongoing data requirements.

Remember, the core challenge of data projects is moving from ad hoc habits to intentional strategies that support an ongoing supply chain. Data as a service requires continuous support and maintenance that must be accounted for in every business stakeholder's strategy. The data team and executive sponsor drive the organisation's data strategy, but that does not happen in a vacuum. Each business stakeholder, through their set of ongoing needs at each point of the data supply chain, provides inputs to that overarching data strategy – and the act of supplying requirements should happen within the communication rhythm outlined in the project materials.

For example, data strategies tend to be driven by summarising all the ad hoc reporting requirements from the business, understanding what platforms and systems support those requirements and, from there, beginning to get a picture of the data supply chain (Plan). As we have described up to this point, the project lead brings the working group and sponsors together and gains agreement on the first pass at critical success factors, critical success criteria and key performance indicators that drive the business (Scoping). Identifying key measures drives prioritisation of reporting needs across the business, likely delivered through a self-service reporting layer (Discovery). To support the reporting layer, data platforms further down in the supply chain must be identified to support the instrumentation, acquisition, storage and transformation efforts for these measures (Build). The data strategy essentially identifies the data stack that will support the organisation in making sound decisions – *and every business stakeholder needs to understand the shared cost of those asks and fund the data team accordingly.*

Data is not cheap. Running numbers just because we can is not a strategy.

Strategy must be intentional. Data must drive business decision-making. Business decision-making drives values such as efficiency, customer delight, optimisation, cost savings, quality, safety, trust, security, privacy and so on.

Ensuring a successful implementation

The kick-off phase helped identify key elements to ensure that the deliverable gets utilised correctly and continues to benefit the organisation after the project winds down. Now is the time to review that list and implement these strategies – scheduling training sessions and sharing information with key working groups, SMEs or sponsors (or all of them!), preparing them to make the most of the deliverable. The job of the executive sponsor and the sponsor group is to make the people, systems and information required for a successful implementation available to the project lead.

The project lead coordinates and directs project resources to meet the objectives of the project plan. They direct the work, monitor the results, keep track of team performance and stakeholder engagement and carefully monitor and control processes to ensure the final deliverable meets the original success criteria.

Once the data project is delivered, changes are expensive. Changes, updates and new requirements now go through a regular prioritisation of the triple constraint (time, cost, quality). Returning to the example of a simple data strategy, the forum for ongoing review can easily be accommodated in data governance. If data governance does not yet exist, an ongoing review of a business key performance indicator usage and need (from the original strategy) provides a perfect agenda topic and meets your ongoing communication objective from the scoping process.

5. EVALUATE AND ACKNOWLEDGE

After the project moves on to ongoing sustainment, the initial work needs to be closed out. Evaluation and Acknowledgement are often overlooked by many teams practicing more traditional project methodologies (PMP, 6Sigma, Waterfall, etc.) but appear to have an integral presence in more adaptive methods (Agile, Scrum, Kanban, Lean, etc.). This is probably because shorter sprint cycles present lower risks than release cycles of 18–24 months, where the gamble of a release lasts much longer. Shorter release cycles require more frequent communication and are more open to data-driven change than traditional methods, all of which lends itself to ongoing learning.

This guide avoids adhering to any particular method. So few data projects craft a scope so precise as to not iterate throughout the process, and there is often a 'fine tuning' aspect to implementation and scale, which we'll cover in the next chapter. Organisations no longer adhere to a singular approach as was the case in the 1980s. Today, people across the organisation share a variety of certifications. Project methodologies are like religion. Everyone has an affinity for one over the other and de-amplifies the dogma to gain the consensus needed for progress. This guide attempts to synthesise the important aspects of getting through the process. It leaves the amplification or de-amplification of rigour to the creativity and expertise of the project lead and what the organisation's culture allows.

Post-implementation review

The main difference between a pre- and a post-implementation review is when the meeting occurs. The pre-implementation review happens before a project begins, and the team proactively considers everything that could go wrong as a way to mitigate risk. The post-implementation review occurs after the project has been implemented and is a time for the group to reflect on the things that happened during the project that went well or might be improved. A post-implementation review is:

1. A method used to gather feedback from a team after a common work effort to discuss and make improvements in the future from lessons learned.

2. Best if conducted soon after work is complete.

3. Can be used with teams of any size, but the method may need to be adjusted for large groups.

4. Considered a best practice in nearly all project and continuous improvement methodologies.

Generally, it is good practice to have a peer project manager who is impartial and uninvolved in the project conduct the post-implementation review. Many project leads wind up leading their own post-implementation reviews for many reasons (i.e. time, convenience, etc.) and there are many ways to still collect anonymous feedback. At first it might make everyone feel a little queasy, but if thoughtfully and respectfully facilitated, the meeting can be an extremely fruitful dialogue (and not a gossip session or forum for personal attacks, as most people fear).

Regardless of who facilitates, the meeting should involve all the key players/groups from the project. The goals of the post-implementation review should be communicated in advance. (What do you want to learn from the exercise?) People should not use the time to go through the motions because it is required.

The facilitator will lay the ground rules. Feedback should be specific and constructive; everyone should 'check their ego at the door'. Feedback is then categorised and ranked in importance. The project lead will not have time to act on everything, so the group needs to understand what is most important. The facilitator then summarises all feedback, distributes it to participants and interested parties and posts it for later reference.

The project lead can then create a reasonable action plan for ongoing reviews, and assign ownership. It is an opportunity for everyone to think more broadly than just this project, feature or milestone – everyone can take learnings from this and apply them to other work. There might be deeper opportunities for alignment or new areas for improvement horizontally or vertically in the organisation.

Before closing the project officially:

- Define an evaluation criteria for the project that will also be used in the post implementation review. What are your goals, and what measures will get you there?

- Celebrate the dedication and impact of your data team, stakeholders and sponsors. No change effort happens because of a singular effort. Acknowledging that goes a long way towards sustaining investment, keeping up belief and enthusiasm for the ongoing work and preparing people for the next project.

Evaluating the success of the project

Each project is a learning opportunity. If you make the space for the people involved and share feedback, then ask the right questions, the project lead can learn important lessons about how your organisation works and what your future data project efforts should look like.

As with all kinds of evaluation, efforts to understand the impact of a data project will look different each time. What are you trying to learn? The answer to this question should inform how you structure your evaluation. Some common needs are to understand:

1. the sponsor and stakeholders' satisfaction with the overall process;
2. the ability of the deliverable or engagement result to meet the stated need and project scope;
3. the ability of the deliverable or engagement result, after [specified time], to have demonstrated an impact on the stated need;
4. the ultimate impact of the deliverable or engagement result to help the project lead have an impact on their mission and stated goals.

See Phase Four: Tune the Change to read more about evaluating data projects.

Collecting feedback

There are many ways to collect feedback, including third-party interviews or qualitative and quantitative surveys. Two common methods are:

One-on-one interviews: Schedule a dedicated time to sit down with each stakeholder and ask specific, focused questions. This more intimate, conversational setting can help you gather more information on particular areas of interest beyond what a consultant might be compelled to report on through survey questions. Challenges include the possibility that team members might provide opaque or overly optimistic answers (that is, 'sugarcoating'). There can be several reasons for this: for example, they may not have the skills to provide direct feedback, or speaking indirectly might be preferred in their culture.

Listening sessions: An efficient way to gather a range of opinions and provide a sense of closure from the team, holding single or group feedback sessions can help foster effective dialogue between team members and stakeholders. Positioning this type of evaluation setting as an opportunity for learning and a time to determine how to improve future data projects is critical. Team members may be reluctant to provide candid feedback in some settings, so ensuring facilitators know that the lead wants to hear all feedback – the good, the bad and the ugly – is very important. Challenges include the possibility that team members are not transparent in front of their peers, or you could encounter 'groupthink' when the team members influence each other to present a single, unified perspective instead of individual points of view.

- Design this conversation to run like a focus group.
- Prepare five to seven open-ended, broad questions and a few prompts for each in case the group gets stuck.
- Establish ground rules (for example, 'We want to hear all of your feedback – good and bad – and we encourage you to respond directly to each other.'). The facilitator guides the conversation, but their primary role should be observing, listening to and capturing the dialogue, not running it entirely. They should interject primarily to synthesis and paraphrase.

However the information is collected, creating space for two-way dialogue is important. The point is to understand everyone's experience, their satisfaction with the group's interactions and how future efforts can improve. Exhibit 5.5 illustrates an example post-implementation review from a group discussion. See Appendix E for review templates.

In the example, the 'mostly good' and 'mostly issues' columns offer a summary of statements that map to verbatims from a group feedback session (located at the bottom of the table), where statements were grouped into themes. Verbatims from one-on-one interviews will be collected and summarised in the same manner. Notice how the direct comments were summarised into themes and ranked and how the group generated recommendations. These issues are so common that you may have experienced many of them yourself.

Exhibit 5.5 Example post-implementation review from a group discussion

Mostly good	Mostly issues
Brainstorm positives, wins and acknowledgements from the team	Issues are clustered below with context

Mostly good

Brainstorm positives, wins and acknowledgements from the team

- *Increased awareness of the importance of business cases and risk assessments, especially as an objective measure for prioritisation – resulted in piloting business cases with robust cost-benefit estimation*
- *Use of a shared website as the single repository for project docs increased much-needed transparency/built trust*
- *Division-wide IT needs were considered for prioritisation (but the execution had issues listed)*
- *Highlighted lack of enterprise-wide data architecture as a critical gap, resulting in the draft architecture for the current strategy exercise*

Mostly issues

Issues are clustered below with context

- *IT-business partnership, process management and accountabilities are not clear*
- *Internal alignment on 'What is on the table vs off the table' for prioritisation and timing changed frequently*
  - *In-flight projects were 'out' but changed at the 11th hour to 'in'*
  - *Project alignment to data team vs operations sponsor (name) top priority projects not aligned with data team/ executive sponsor (name)*
  - *Balancing individual priorities vs org 'good' was not facilitated – no way to know how one project stacked up against another*
  - *Core team would not make trade-offs for closing the $12M gap, resulting in 'throwing up hands'*
- *Role of offshore resources was not communicated or clarified, resulting in sub-optimal interactions*
- *New tool was introduced without training*

(Continued)

Exhibit 5.5 (Continued)

Issue area	Recommendations
Issue themes	Brainstorm from the group or from one-on-one interviews

1. *Internal alignment on 'What is on the table vs off the table' for prioritisation and timing changed frequently*	• *Clearly define (and stick to) criteria and policies for project evaluation and categorisation and consistently use it, including:*
	  ▪ *Use of business cases, mapping to common approach*
	  ▪ *Alignment with data team and processes*
	  ▪ *Definition of in-flight, corp mandate business relevance*
	  ▪ *Use of risk assessment*

2. *IT-business partnership, process management and accountabilities unclear*	• *Develop three-year roadmaps* • *Map business architecture efforts to project inventory* • *Continuous portfolio analysis: project values (over the life of the project) assessed against preliminary portfolio regularly vs once a year*
3. *Vendor resource responsibilities weren't explained, leading to less effective interactions*	• *Better integrate, manage and communicate consultant involvement and interactions with existing business and IT partnership structure*
4. *New tool (name) was introduced, no training*	• *Increase training of tool (name) usage* • *Provide documentation*

Issue area	Verbatim issues
Verbatims grouped into issue-themes	Direct phrases from interviews or group dialogue
Internal alignment on 'What is on the table vs off the table' for prioritisation and timing changed frequently	• *Comes down to a better coordination effort – it felt uncoordinated with the frequency of last-minute changes. (It may have been due to fast track)* • *11th hour changed the taxonomy; this is not good* • *Avoid 'false starts and stops'* • *Reduce fire drills* • *Need more explicit guidelines – changed course all the time. Not a transparent approach*

(Continued)

Exhibit 5.5 (Continued)

	• *We need to have well-defined terms and use of the terms. What is in flight needs to be in flight across the board. Business relevance and corporate mandate*
	• *Many projects aligned directly to Ops Sponsor (name) goal but did not align to a core imperative*
	• *There was a poor translation of projects to the data team's 'token goals'*
	• *Needed more clarity on what core imperatives were, etc.*
	• *Core imperatives appeared to be a parallel effort*
	• *Not sure how to work with core imperatives*
2. *IT–business partnership, process management and accountabilities unclear*	• *Ensure business sets the priority*
	• *How do we ensure we continue planning throughout the year, assessing project continuously? Need division perspective for prioritisation*
	• *Business partners asked, 'Why am I getting "randomised"?'*
	• *Had a hard time assigning one single person to a core imperative*
	• *For needs that were unique to one business area, there was no way to get them on the radar for consideration*
	• *Business partners need to be engaged at the beginning of the planning and prioritisation process*
	• *No reference to any of the FY06 planning timelines set out by [major corporate planning event] to that point in time*
	• *Need to allow COO responsibility of working the client relationship*
	• *The client is the pivot-point, not the core imperative – having this backwards pivoted the effort the wrong way*
	• *No way to get business priorities that did not align to core imperatives in front of an audience*

(Continued)

Exhibit 5.5 (Continued)

	• *Simplicity – the process and the data captured should be simple and clear in order to be used*
	• *We need to set out a plan and march to it*
	• *IT still programme-manages the structure for getting through planning*
	• *Documentation was vague*
3. *Vendor resource responsibilities weren't explained, leading to less effective interactions*	• *There was little communication from the division team about the role or reason for bringing in vendor resources*
	• *No plan for incorporating vendor resources' work with ongoing planning with IT and business teams*
	• *Confusion about who was sponsoring resources*
	• *Feedback seemed ignored by vendor resources and division leadership*
	• *Hard work throughout the year felt invalidated due to new interactions by vendor resources*
	• *Missed understanding the business drivers behind the planned activities using three-year roadmaps that were already started*
	• *Missed understanding the daily activities of business groups, not recognising other priorities than what vendor resources had, leading to extensive follow-ups*
	• *Involving a third party made things harder and take longer*
	• *This added time and money disrupting the COO's collaboration with the corporate mid-year planning*
4. *New tool (name) was introduced, no training accompanied it*	• *Simplify the tool and the process it supports*
	• *More business training in the use of the tool; more business and IT training in using the tool over the life of projects and in managing the business sponsor portfolio of projects*

(Continued)

Exhibit 5.5 (Continued)

- *Ensure tool and process align with release management needs*
- *Link tool to business problem and validate ROI through case study*

Looking it over, it might seem like the 'mostly goods' didn't get much space or attention among the group. Much of the 'work' of this meeting is learning from what didn't go well and considering how to exploit that knowledge and transfer it into opportunities.

Sometimes, simply highlighting a 'lack of enterprise-wide data architecture' can go a long way towards increasing awareness, seeding productive dialogue and generating collaboration. Removing such a blind spot or directly addressing a taboo in the organisation could challenge 'the way we have always done things' or directly impact a senior leader's budget and resources – which is why it was being ignored. Now, more people know of the issue, and it's given context to a $12M opportunity to fix – which is impossible to ignore. The other reason to spend so much time teasing apart the things that didn't work is to expose where sponsors might have conflicts or competing commitments. They may or may not be aware of this, and now the sponsor forum can serve as the escalation and decision forum it was meant to be, because the leaders are named.

Key areas for reflection

It can be helpful to plan some reflection time on the overall engagement from multiple angles, to use the data project results for more than the task at hand. The extent of the evaluation performed is tied to available resources and what you want to learn about the work. For a general continuum of evaluation questions, see Exhibit 5.6.

For any project, you should explore the basic issues of how the project went and what you can do as a result. See Exhibit 5.7 for sample questions for self-reflection and Exhibit 5.8 for sample questions for sponsor self-reflection.

Core evaluation criteria

For all projects, no matter what the type of project, the result, the team size or the resources, there are standard criteria against which the success of projects should be evaluated. While the evaluation may cover a full range of intended outcomes, all successful projects should have the following characteristics:

- The project is completed and delivered in a timely manner.
- The deliverable can be implemented and is sustainable.

- The executive sponsor, sponsor group, working group and project lead's expectations are met or exceeded.

- All parties involved report high satisfaction with the group's interactions. (One company goes as far as to ask: Would we want to work together again?)

- The stakeholders (e.g. beneficiaries and extended beneficiaries) report that the project made an impact.

- Recognise that addressing immediate issues reactively and with a short-term perspective is insufficient. We should view today's problems as a chance to anticipate and address future challenges.

Exhibit 5.6 Basic to advanced evaluations

Satisfaction	Success of project		Organisational impact
Satisfaction with the team interactions	**Meeting of stated need**	**Long-term solution**	**Increase capacity to deliver on stated goals**
Did those involved with the process feel satisfied with how the group interacted and the overall process?	Does the final deliverable address the challenge that it was intended to resolve?	Will the impact of the project extend over 12–24 months? How do you expect it to continue to improve the organisation's operations?	Can you demonstrate an increased ability to deliver on your stated charter through the impact of the data project?
Evaluation in this category looks only at how the project went. At a minimum, all projects should be evaluated on these criteria.	Evaluation in these categories looks at what the data team can do as a result of the engagement.		Evaluation in this category looks at the ability of the non-profit to deliver impact on its stated mission and goals because of data project work.

Exhibit 5.7 Sample questions: project lead experience

Use the questions below to prompt a draft of your own evaluation questions.

Satisfaction with the team interactions	Meeting of stated need	Long-term solution	Increase capacity to deliver on stated goals
• What was the most beneficial part of the project? The most challenging part? • Were sufficient processes in place to keep the team engaged and committed throughout the process? If no, please describe. • How would you rate the team's interactions? Do you feel the dynamics fostered effective collaboration? Why or why not? • Did you feel the scope adequately prepared you for the expected contribution? • Did you feel the project lead was responsive, respectful, and open to mutual learning? • Now that project work has concluded, do you plan to stay involved with your data team partners? If yes, how?	• Did you feel you could deliver what was expected from you through the provided time and scope of the project? • What will your extended stakeholders be able to do now that they could not before? • How have you prepared the stakeholders to implement and sustain the deliverable? • What additional follow-up work would strengthen the deliverable or allow the stakeholders to take next steps in thinking through requirements or the data team to build on the solution provided?	• Did you feel you could deliver what was expected from you through the provided time and scope of the engagement? (~18 months after the project is delivered) • What will the stakeholders be able to do that they were unable to do before? • How have you prepared the stakeholders to implement and sustain the deliverable? • What additional follow-up work would strengthen the deliverable or allow the stakeholders to take next steps in thinking through requirements or the data team to build on the solution provided?	• From your perspective, how does the data project help the stakeholder meet their stated strategic priorities? • What indicators will help track if the project is changing or improving? • How do you suggest that the client track these indicators? What systems might already be in use that could be adapted to monitor progress? • How often should indicators be measured/ monitored?

Exhibit 5.8 Sample questions: executive sponsor self-reflection

Use the questions below to prompt a draft of your own evaluation questions.

Satisfaction with the team interactions	Meeting of stated need	Long-term solution	Increase capacity to deliver on stated goals
• What was the single most frustrating part of the project? What was the most beneficial part of the project?	• What can you do now that the work is complete that you could not do before? (right after the project is delivered)	• What can you do now that the work is complete that you could not do before? (~18 months after the project is delivered)	• Which strategic priority were you hoping to address through this data project?
• How should we do things differently to avoid frustration?	• How will the new deliverable or learning help you meet the stated need?	• How will the new deliverable or learning help you meet the stated need?	• What indicators do you use to measure your progress on this strategic priority?
• Which of our methods, processes or steps worked particularly well? Which performed particularly poorly?	• What challenges do you expect in making the most of the deliverable?	• What challenges do you expect in making the most of the deliverable?	• Using those indicators, can you correlate a change in your progress related to the data project?
• Were the stakeholders and sponsors sufficiently thanked?	• How long do you expect the deliverable to be used? How will you govern (determine ongoing requirements, think through ethics of use, drive policy/compliance) the deliverable?	• How long do you expect the deliverable to be used? At what point will you need to update or revise the deliverable?	• If you can afford a third-party audit, have them assist you in tracking and monitoring your impact (and overall data management maturity).
• Was meaningful and actionable input provided to the data team? To stakeholders? To sponsors? What are some examples?	• What will tell you if this project has been successful in 12–18 months?	• What did you select as indicators when the project was first completed to track the success of the work? What do those indicators suggest?	

Acknowledging the team

The core components of an effective working relationship are trust, reciprocity and exchange of value. These qualities, particularly the latter, leave our working partnerships happy, rewarded and satisfied with the process. Being intentional about how we acknowledge others has essential benefits:

- Increased customer satisfaction. There is a direct and positive relationship between employee satisfaction and customer satisfaction; and one of the main factors in improving employee satisfaction is recognition.[77]

- Reduced negative attrition. Acknowledging people's efforts contributes to them feeling valued and increases their desire to continue working with you. According to Gallup, employees who don't feel recognised are twice as likely to want to quit in the following year.[78]

- Reduced absenteeism. Employees tend to take less sick leave or days off when they are reminded of how important they are to their organisation. Gallup research shows that if businesses double the number of employees they recognise every week, there will be a 27 per cent reduction in absenteeism.[79]

- Increased productivity. Employees who feel recognised for their work are more likely to put in more effort and work harder to be recognised again. Research from Deloitte shows that employee productivity and engagement are 14 per cent higher in organisations that practice acknowledgement.[80]

Some leaders feel inauthentic or uncomfortable sharing recognition and gratitude, for several reasons. The sentiments below are paraphrases from leaders who feel challenged in giving credit. Leaders who withhold recognition do so because of deeply entrenched beliefs. Whether it is conscious or subconscious, leaders often withhold appreciation because they believe one or more of the following:

- In the business or in the bigger picture, showing appreciation is not a priority.

- No one thanks me, so why should I go out of my way to praise others?

- I don't want to praise good work from someone where other areas need work – it may seem like I condone the latter.

- I have extremely high standards; people must over-deliver to get my attention.

The impact of our beliefs is profound. Beliefs guide our behaviours. Ultimately, what we believe (and act upon) can block appreciation and multiple forms of progress. Our project might run slower, our relationships might take longer to build and our resilience in working with others gets depleted.

We take our cues from leaders (that might be us!) and consciously or unconsciously mimic their behaviours because that is what is being modelled. A single leader can cultivate an unappreciative cultural effect. Negative thinking overshadows the positive; from there, it is like dominos falling: our thinking becomes constrained and our morale declines, which informs our beliefs and creates a good environment for groupthink.

Key takeaway? We matter, whether we like it or not.

---

77 https://www.researchgate.net/publication/332410189_Employee_Satisfaction_and_Customer_Satisfaction_-A_Close_ Comparison

78 https://www.gallup.com/workplace/236441/employee-recognition-low-cost-high-impact.aspx

79 https://www.gallup.com/workplace/238085/state-american-workplace-report-2017.aspx

80 https://www2.deloitte.com/content/dam/Deloitte/mx/Documents/about-deloitte/Talent2020_Employee-Perspective.pdf

Bringing awareness to negative emotions we have around acknowledgement – we have all probably felt at least one – highlights that we need to practice giving acknowledgement. Recognising what we want to reinforce must feel genuine to us, first, to land authentically with the recipient. Table 5.22 shows how we can constructively move from judgement to practicing acknowledgement in small ways.

Table 5.22 Practising constructive acknowledgement

Situation	Emotion	Belief(s)	Mitigation
In a project update, a stakeholder finally delivered their portion of the contribution despite organisational chaos and demands. They had been lagging throughout the engagement.	Frustration, resentment.	'I don't want to praise good work from someone where other areas need work – it may seem like I condone the latter.' 'I have extremely high standards; people must over-deliver to get my attention.'	Take time to acknowledge minor improvements, especially if they highlight a before/after impact. For example, this person's contribution was lagging, but they delivered. When something we've done is recognised ('It was so helpful to the team that you handled that'), or, even better, someone acknowledges an aspect of who we are ('I appreciate how you handle tough situations so diplomatically'), first and foremost, we feel seen.

Why the need to shift this withholding habit to one of greater appreciation? Because acknowledgement does the following:

- It tells us we belong. We may or may not remember the last time someone complimented us in a meeting, but it likely felt good. Even a small but sincere 'Great point!' (if it doesn't get overused) says we added value and stokes a sense of belonging – the need for which is hard-wired into our DNA.[81]

- It fuels purpose, an intrinsic motivator. Acknowledging someone's impact can affirm that we're contributing to something bigger than ourselves. Intrinsic motivators come down to learning, autonomy and a sense of purpose.[82]

- Giving thanks and being grateful feels good for everyone. Showing appreciation to someone is like giving a gift. Both givers and receivers of a gift experience a jolt of the 'feel-good' hormone, oxytocin.[83] Conversely, cortisol is released when we're given negative feedback. As leaders, we need to receive and provide *both* types of input, so delivery matters. Our brains have a negative bias; we attach to

81 https://hbr.org/2019/12/the-value-of-belonging-at-work

82 D.H. Pink, *Drive: The Surprising Truth about What Motivates Us* (Riverhead Books 2009).
Note: Pink's treatment of autonomy is very US-centric and does not take into account other cultures' need for consensus, or the individual–group relationship. However, acknowledgement does reinforce our sense of purpose.

83 https://www.whartonhealthcare.org/discovering_the_health

the negative to stay alert and survive (the jungle, office politics, etc.). It takes five positive statements to counter one negative statement.[84]

- It contributes to a positive work culture. Each one of us is a culture. How we show up (or don't), trust (or don't), and engage in constructive norms (or don't) is culture. Building in team rituals that include acts of appreciation reaps all of the above benefits, with the bonus of nurturing engagement, social safety and enthusiasm. Acknowledgement creates collective energy that is hard to come by if we leave it to chance.

How you choose to recognise people is unique to you. It is especially meaningful to learn how others uniquely wish to be appreciated. It's easy to assume that others want to be appreciated in the same way that we do – yet, for example, highlighting someone's good work in a big meeting may be great for one person but mortifying for another.

Phase 5: Evaluate and acknowledge checklist

Before considering the project closed, double-check that you have completed the following:

1. Key goals for evaluation have been outlined and vetted with stakeholders.
2. An evaluation process has been planned, including preparing survey questions or discussion guides.
3. All members (data team, any offshore resources, stakeholders and sponsors) have completed the evaluation.
4. All members have been thanked and plans have been made for continued recognition (for example, year-end updates).
5. Appropriate stakeholders have seen the results of the project work and have been notified that the project has been completed.
6. External stakeholders have been notified that the project was completed successfully, including offering to share a quote or case study write-up.
7. A plan for ongoing adjustments has been determined.
8. The group has gone through an 'ethics fast-check' review to provide a forum for concerns and document mitigations.
9. Determined how to acknowledge team members, stakeholders and sponsors.

KEY POINTS

- Building foundational skills through managing easy or intermediate projects prepares you to transition to more advanced data work, like weighing data privacy with data democratisation initiatives or thinking more deeply about the ethics of data storage or the sharing or selling of that data to third parties.

---

84 E. Lisitsa (2012, 3 December). The Positive Perspective: Dr. Gottman's Magic Ratio! http://www.gottmanblog. com/2012/12/the-positive-perspective-dr-gottmans.html

- Most mistakes happen because a problem was oversimplified, or where a problem is solved independently by one department or function and applied to all. Looking at complexity solely through the lens of just one function or discipline creates a blind spot.

- Five general principles can take on particular importance for data projects: (1) prepare to sustain your effort from the start; (2) embody your stakeholders; (3) communicate, communicate, communicate; (4) engage in two-way learning, always; and (5) recognise goodness – be behaviourally specific.

- The planning phase clarifies the primary objectives, establishes a good rapport with the team, ensures the team is familiar with the mission and has reciprocal alignment to the data project's goals. Ensuring early support paves the way for a smooth kick-off. Specifically: does the data strategy support the business strategy?

- Fostering the kind of relationship that makes lives easier and ensures great work requires establishing trust, cultivating mutual receptivity and providing value.

- The data project lead's role is to amplify the exchange or value for their stakeholders. Never miss an opportunity to connect the work to the end impact you expect to see. When this is done with authenticity, it's like handing peers and stakeholders a tiny bonus.

- Kick-off isn't typically a phase, but many skip past opportunities to consider meeting details and design. Kick-off focuses on preparing for the first meeting, bringing together the internal data team and the working and sponsor groups. Keep a service mindset throughout the project – data projects are part of a sustained service.

- Discovery helps to better identify project scope and goals, resulting in more accurate budget and timeline estimates. It is designed to help the team learn as much as possible about the problem. Discovery enables a strong, collective understanding of the organisation's needs – both generally and related to the data project.

- The Build phase is about creating, developing, improving, writing or producing the end process, procedure or solution. This phase looks different for every deliverable; there is no step-by-step process, but there are some best practices to consider to be successful.

- The final group meeting is the 'official' handoff of the deliverable, and it is important to bring the whole working group and sponsor group together for this presentation. The groups have seen and reviewed iterations of the deliverable before this point, so they have a good sense of what the project lead will present. If there are any surprises in this meeting, the deliverable is not delivered; you are still in the Build phase.

- After the project, before it moves on to ongoing sustainment, the initial work needs to be closed out. Evaluation and acknowledgement are overlooked by many teams practising more traditional project methodologies (PMP, 6Sigma, Waterfall, etc.) but appear to have an integral presence in more adaptive methods (Agile, Scrum, Kanban, Lean, etc.). This is probably because the latter are quicker (results in weeks versus months or years), require more frequent communication and are more open to data-driven change than traditional methods.

- The main difference between a pre- and a post-implementation review is when the meeting occurs. The pre-implementation review happens before a project begins, and the team proactively considers everything that could go wrong as a way to mitigate risk. The post-implementation review occurs after the project has been implemented and is a time for the group to reflect on the things that happened during the project that went well or might be improved.

- Some leaders feel inauthentic or uncomfortable sharing recognition and gratitude for several reasons. Leaders who withhold recognition do so because of deeply entrenched beliefs. The impact of such beliefs is profound. Ultimately, what we believe (and act upon) can block appreciation and multiple forms of progress. Our project might run slower, our relationships might take longer to build and our resilience in working with others gets depleted.

6   TUNE THE CHANGE

Now that we've reviewed the method for handling a data project from planning to execution, how can we extend this approach across the entire organisation? As mentioned in the Introduction, transformative work involves ongoing change, akin to tuning into the right frequency. Managing continuous change requires collective support from the entire organisation to ensure the success of each endeavour.

In this chapter, we'll explore:

- five characteristics essential for an effective data organisation that you can cultivate over time (Table 6.1);
- strategies you can develop to assess the necessary infrastructure for successfully handling a growing number of data projects;
- methods for fostering strong relationships to guarantee the long-term success and sustainability of your organisation's data initiatives.

BECOMING DATA-DRIVEN

Before moving to the specifics of how to become data-driven, focus on why moving from an ad hoc effort to a mature approach to becoming data-driven makes sense. Remember the maturity curve? The timing might not be right (now). Becoming data-driven through data as a service requires a serious investment of resources – financial, staff, equipment, services, etc. – and scaling your efforts will only increase those topline demands.

Completing a data project (or several) has shown you the benefits of having a repeatable approach that tends to the human and machine side of the project. Much can be done to hone the approach you've developed. Similarly, the benefits of scaling include:

- Accomplishing more with less.
- Fully engaging executive sponsorship to tap into a deeper, more interconnected form of strategic planning (one where data is intentionally incorporated into each business's goals).
- Cultivating increased support for data-driven decision-making by exposing extended stakeholders to tools, resources and support.
- Addressing challenges and obstacles that would otherwise be difficult to overcome.
- Cost savings from consolidating the highest-priority process, platforms and data centrally, with federated efforts in each business unit.

Figure 6.1 illustrates an organisation's key skills and behaviours driven by data.

We can observe the following characteristics in a data-driven organisation:

- Data is used across every function to analyse decisions, measures and best practices continuously.

- Data is a core component of every business unit's strategy.

- Each business unit proactively shares its strategy with the data team so that the data team can plan a common architecture and set of tools and services. The data strategy is aligned or tightly integrated with the IT strategy.

- Related to the above, leaders foster strong collaboration between data and business teams.

- Leaders model data-driven decision-making transparently and hold their teams accountable for the same behaviours.

- Data teams enable the optimisation of resource allocation.

- Ongoing issues, risks, strategy updates and policies are reviewed in the governance forum.

- A broad data mindset is cultivated by both human resources (that is, recruiting for data skills, competencies, performance management and general digital upskilling) and data teams (integrated data strategy, data services and enablement) by expecting every employee to model effective decision-making in every meeting.

- Data instrumentation and collection standards are a primary concern.

- Human resources is a primary strategic partner and infuses existing mechanisms with data-centric language, behaviours and training (for example, corporate values, performance management programmes, leadership development programmes and so on).

SHIFTING CULTURE TO BE DATA-CENTRIC

Moving from an ad hoc data provider or consumer to operationalised efforts requires people, processes and platforms to align their thinking about and approach to data projects. Data has legitimately become a fourth pillar in the equation. Table 6.1 presents five characteristics required to tune into the right change frequency. These efforts won't happen overnight. Cultural shifts can take years. That is why diagnostic and maturity curves depicting each phase as equal (as mentioned in Chapter 1) are incredibly misleading.

An organisation could be at level two or three for quite some time. Happily, the pace of change depends on the business pressures to move from one level to another and healthy internal mechanisms to help enable that shift. The following section outlines a strategic approach to tuning into the right frequency and explains when a cultural change might be considered.

Figure 6.1 Inside the data-driven organisation

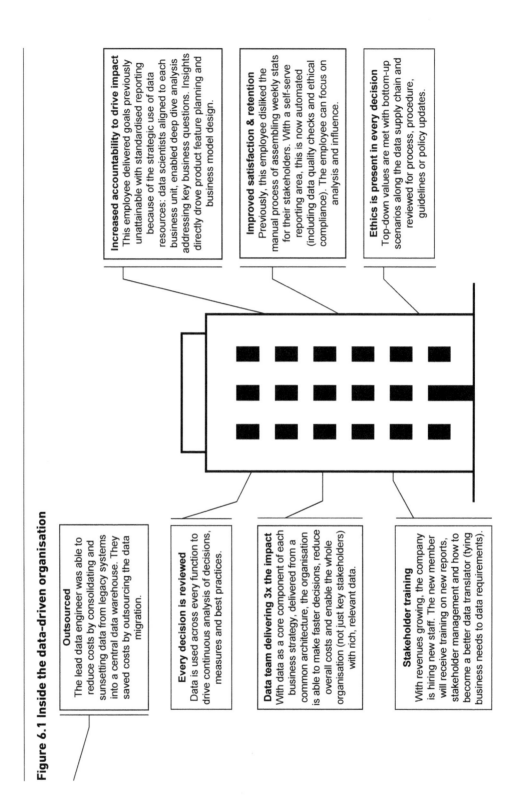

Outsourced
The lead data engineer was able to reduce costs by consolidating and sunsetting data from legacy systems into a central data warehouse. They saved costs by outsourcing the data migration.

Every decision is reviewed
Data is used across every function to drive continuous analysis of decisions, measures and best practices.

Data team delivering 3x the impact
With data as a core component of each business strategy, delivered from a common architecture, the organisation is able to make faster decisions, reduce overall costs and enable the whole organisation (not just key stakeholders) with rich, relevant data.

Stakeholder training
With revenues growing, the company is hiring new staff. The new member will receive training on new reports, stakeholder management and how to become a better data translator (tying business needs to data requirements).

Increased accountability to drive impact
This employee delivered goals previously unattainable with standardised reporting because of the strategic use of data resources: data scientists aligned to each business unit, enabled deep dive analysis addressing key business questions. Insights directly drove product feature planning and business model design.

Improved satisfaction & retention
Previously, this employee disliked the manual process of assembling weekly stats for their stakeholders. With a self-serve reporting area, this is now automated (including data quality checks and ethical compliance). The employee can focus on analysis and influence.

Ethics is present in every decision
Top-down values are met with bottom-up scenarios along the data supply chain and reviewed for process, procedure, guidelines or policy updates.

Table 6.1 Characteristics of a data-driven organisation

Forecasting data needs	• Constant look ahead at 6–12 months out, to identify areas of need that can be addressed through operationalising data requirements.
	• Thoughtful, vetted list of data projects ready to map to existing resources or quantify headcount requests.
Developing core systems to ensure quality and consistency	• Leverage existing systems to manage all aspects of the data supply chain. Track what is inside and outside the commonly sponsored architecture and determine if/how it should be included.
	• Ensure every member of the data team and larger stakeholder group feels valued and appreciated through planned and unplanned activities (team meetings, project meetings, morale events, etc.).
	• Ensure all members are trained on data supply chain basics. They should be able to teach others and help evangelise the value of data-driven activities in the business.
	• Maintain a record of past, current and future projects.
Encouraging deep collaboration between sponsors, working group and data team	• Sponsors, working groups and the data team are true partners in scoping, resourcing and managing data projects.
	• Each has clear roles and responsibilities, and each is armed with the tools and knowledge to be successful.
Building strategic partnerships to support ongoing resource needs	• Strong partnerships with organisations who are in it for the long game.
	• Partnerships fulfil a variety of needs – budget contribution, resource alignment/allocation/sharing, committee participation and sounding boards for advice.
	• Members believe in your mission of integrating data into the business; they understand who you are and what you do and provide holistic support.

(Continued)

Table 6.1 (Continued)

Thinking resource-conscious, not resource-constrained	• Opportunistic about using stakeholder resources, but you'll never forgo the areas of strategic importance that warrant actual topline investment instead. You're not cheap; you're scrappy.
	• For all data projects, you first ask: Is there another way to do this?
	• The starting line-up comes from the data team, but employees from other groups are just one ask away from becoming included. The key is not securing resources; it's figuring out how to get folks included and in the game.

THE TUNING PROCESS

With greater clarity on what your organisation will look like when data-driven, it's time to review the steps you'll take to get there.

It is important not to allow perfection to become the enemy of good. Perfection isn't achievable in any environment, and good is way better than nothing. Short sprints with regular releases and constant customer/stakeholder feedback will beat long-cycle deliveries every time. Many companies still lack that 'customer-first' mindset and/or the ability to operate in short cycles, which would keep their stakeholders engaged and help them improve their data assets more rapidly.

Committing to data as a service means committing to ongoing organisational transformation. Consultants write about 'transformation', 'shifting' and 'pivoting', which implies that there are firm coordinates from point A to point B. However, common external pressures (for example, digital and emerging technologies, changing market conditions, regulatory pressures) and internal requirements to meet the changing demands (for example, new leadership; mergers or acquisitions) that spark change require ongoing alignment – much like tuning into a signal frequency. Increasing data interdependency with business strategy means constantly aligning your organisational structure, talent, leadership and culture with your business strategy to sustain change and keep pace over time. That is why seeing a path through the forest and knowing where you can expect to be as you develop is important.

Everyday external pressures and internal requirements to meet the changing demands that spark change require ongoing alignment – much like tuning into a signal frequency.

In the coming sections, we will review a roadmap for scaling based on a simple maturity scale and what is needed to pass from one level to the next. There are no step-by-step instructions – every organisation is at a different stage, with unique needs requiring unique adjustments. How fast you move through each level depends on many factors, some of which you have experienced while driving your data project (adequate sponsorship, reasonable budget, resourcing, strong desire for progress, the right skills and capabilities and so on). Figure 6.2 shows the PCT Model again, emphasising the integration objectives from both project and change management.

Figure 6.2 ProSci PCT Model (Source: ProSci, PCT and 'Project Change Triangle' are trademarks of ProSci Inc. Copyright © ProSci Inc. All rights reserved.)

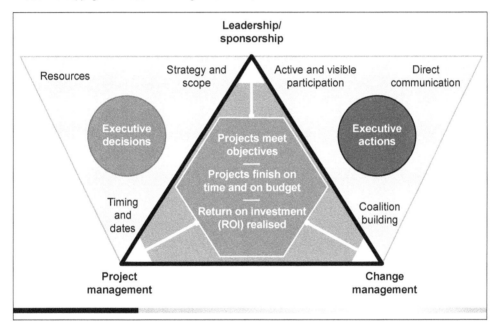

Whether you follow these steps or develop your plan for expanding current data efforts, work with and towards the five guiding principles of successful data teams mentioned in Phase Three: Manage the Work. Moving from level to level will require a discrete change effort. Ultimately, the group you identify grants permission to move from one level to the next. They must be aware of the need to get to the next level, desire to get there, knowledge of what needs to happen for the shift to take place, ability to make the shift happen and constant reinforcement of the shift by project leads and sponsors in order to maintain it.[85]

---

[85] ProSci's ADKAR model provides an excellent framework for many types of organisational change.

The tuning process outlined in Table 6.2 represents the easiest way to get from the start to the endpoint, with the lowest chance of falling into the traps of scale. In each section a best practices table focuses attention on one variable at a time. For example, instead of increasing the volume of projects while working with a new department within your organisation, choose one or the other. Incremental steps help highlight one notable change at a time, which will help preserve quality and maintain forward progress.

Table 6.2 The tuning process

Observer	Practitioner	Learner	Skilled
Exploring the use of data and gaining a basic understanding of the data supply chain	Completing a few data projects in a single functional area, such as marketing or engineering	Expanding the use of data projects through a succinct initiative, spanning multiple departments, programmes and operational staff	Using data across programmes, operations and finance and leveraging insights as a core component of a shared strategic plan

LEVEL ONE: OBSERVER

All organisations start here. As new paradigms for the future come into view, we begin to observe what needs to be done. We develop an understanding of the things that need to shift. As mentioned in Chapter 1, most organisations – even large, sophisticated companies – engage with data on a highly ad hoc basis. There is a general awareness that data is important, so there is motivation and interest to put something together that helps drive decisions for the group, maybe even influencing close stakeholders. Yet, there is difficulty in completing projects larger than ad hoc reports and even more resistance to automating those efforts. Motivation is high, but people burn out using current mindsets and tools as demand increases.

Moving to the next level requires fostering a mindset for change. Implementing a few best practices and processes from Table 6.3 can boost confidence in enhancing stakeholder experience. Even in this initial stage, embracing new changes can energise the team and lay the groundwork for a more robust data culture. These early data projects also offer an excellent opportunity to practice effective project, programme and change management. As we progress along the maturity curve, the speed and importance of initiatives will only increase and a lack of foundational skills can pose a significant risk to the organisation.

Table 6.3 Best practices: observer

What roles does the executive sponsor of the data team play?	At this level, the data team and project lead will manage a group of data projects. Additional sponsors will be involved in scoping work or securing resources, but the data team lead will do most of the work.
	The executive sponsor can highlight the value of this work to peers and begin a slow advocacy campaign. Recall early discovery efforts from Phase Two: Determine Resources, specifically the answers to discovery questions for an executive sponsor, and integrate that information into the project planning process.
Which culture shifts should you focus on?	Initial awareness starts with the data team's mission, vision and values. This language matters as it serves as the compass for the group and determines how the work will get done. The first few early projects set the tone for how data work will proceed. This is where the seeds of 'data revolution' are planted.
	Each 'level' emphasises activities where mindsets and small culture shifts are necessary to establish the use of data as an ongoing service:
	✓ Establish mission and guiding principles for the team.
	☐ Forecast and prioritise data needs.
	☐ Develop core systems to ensure quality and consistency.
	☐ Encourage deep collaboration between sponsors and working groups.
	☐ Build strategic partnerships to support ongoing resource needs.
	☐ Adopt a 'resource-conscious' approach, not 'resource-constrained'.
	☐ Constantly review ethics, compliance and standards of data strategy.

(Continued)

Table 6.3 (Continued)

How will you know if you're ready to move to the next level?	After completing your first data project successfully, you are ready to move to the next level. Success is relative, but should include:
	☐ Completing all the phases of the project as defined in this guide.
	☐ Ensuring feedback was taken throughout the process.
	☐ Encountering and successfully navigating unique issues of your organisation that were *not* outlined in this guide.
	☐ You accomplished what you set out to achieve; the general sentiment was positive.
	☐ There is measurable impact from what was delivered.

Creating a mission sounds like a simple thing to do. But when teams are moving quickly, many skip this step and wonder later why they are three degrees off course and feel misaligned. Sometimes, misalignment occurs because the mission is too high-level and vague, making it difficult to translate into concrete initiatives. Teams become confused and focus on 'busy work' rather than impactful outcomes. To create change, data teams looking to seed the revolution must reshape their project portfolios into clearly defined missions directly linked to organisational goals. The mission statement helps support scoping efforts.

The most effective missions incorporate these best practices:

- *Clear and strategically grounded*: The team should be clear about the mission and how it supports the business's overall strategy.

- *Valuable and measurable*: Missions focus on outcomes whose value is measurable, not output or activities.

- *Holistic in scope*: The mission should be broad enough to have a real impact. Teams therefore need to have the mandate and obligation to solve problems across organisational siloes.

Use Table 6.4 to brainstorm and review corporate values as a team. Choose the three that best support a data-forward culture. There are times when some corporate values will fall short. The gap in actual versus aspired-to values, however big or small the change (in this case, driving a data culture), will be new to everyone. Starting within your team becomes even more critical – it is the seat of the data revolution! You will be modelling new ways of thinking, problem-solving and working together – behaviours you want others to adopt. If guiding principles (values, competencies and so on) from your organisation are not descriptive enough (there are some cultures where these are not well defined), look to data-forward companies for inspiration.

Table 6.4 Applied practice: developing a mission statement

Point your compass in the right direction as early as possible. As a team, decide the one to three things your strategy needs to accomplish and succinctly summarise them into one sentence.	
Proposed	**Example**
The _____ mission is to (initiative's or team name)	*The data team's mission is to*
bring data to the _____ (your target) _____, and (goal 1) _____. (goal 2)	*bring data to existing business units, make data-driven decision-making a reality, and* *drive higher performance levels through a standard set of processes, data and systems.*
Now you try: Use Worksheet 13 and complete the first table, 'Developing a mission statement'.	

Guiding principles will be useful when the team hits inevitable setbacks and needs guidance to get beyond them – for example, where the team is analysing data and debating a way forward. There are times when more debate will not change people's minds. Leadership principles that support the mission can offer guidance – at some point, the team will have to 'disagree and commit'.[86] That same conversation might be considering a build-or-buy debate, where a simple guiding principle might solve the

---

[86] Amazon defines this principle as follows: 'Leaders are obligated to respectfully challenge decisions when they disagree, even when doing so is uncomfortable or exhausting. Leaders have conviction and are tenacious. They do not compromise for the sake of social cohesion. Once a decision is determined, they commit wholly.' https://www.amazon.jobs/content/en/our-workplace/leadership-principles

Table 6.5 Applied practice: determining guiding principles

As a team, choose three to five guiding principles from your organisation or inspired by outside organisations. Keep track of how they work and don't work throughout the process. By observing what does and does not work, you are starting the process of right-fitting what data-forward looks like for your organisation and making it easier for others to adopt.

Proposed guiding principles	What works	What does not work
Disagree and commit	This helped move conversations forward.	Need to not use it as a way to shut others down. It can be overused.
Think differently with meaning	This has helped increase the depth of our brainstorming and problem-solving. People are taking time to think through their contributions before bringing something to the team.	
Own technology behind our products and solutions	Initially, this was a value we wanted to have.	As our efforts are still early, we might need to buy a solution to become more educated on what we want to build internally.

Now you try: Use Worksheet 13 and complete the second table 'Determining guiding principles'.

problem with 'We own the technology behind our products'.[87] When the team is trying to problem-solve, choosing a value that validates thinking differently but offers guidance to 'think differently with meaning' can help separate constructive and unconstructive debate.[88] Use Table 6.5 to brainstorm and review guiding principles for your team. When reviewing guiding principles, choose carefully. When using guiding principles, see how they work. Learn and adapt as necessary. Trying on data-forward thinking looks different in every culture.

LEVEL TWO: PRACTITIONER

It sounds counter-intuitive, but ongoing learning comes after practice. Think of lifelong learning as the insights following a sprint release. Sprints are a form of practice that leads to learning and insights. It's the same with leading data projects and managing data as a formal discipline.

After completing a project successfully, consistently tuning ongoing efforts is about deepening practice and honing skills. Driving programmes at scale requires skilled people to emphasise increasing efficiency, reducing costs and driving standards. As with any skill, it is wise to focus on a single function within the data team (programme management, data engineering, data science and so on) and a single model of delivery (for example, starting a project management office, auditing the data warehouse usage and respective costs or building comprehensive data models) before undertaking a completely new data project. Don't try to skip through the levels too quickly.

Sticking with a single practice area means you can cross-train data team members for various tasks, reducing the risk of overdependence and the need to continue training new team members. Doing more projects in the same operational excellence practice builds the team's capabilities. A primary stakeholder from a business unit can also begin contributing to the data team's forecasting – looking at their short- and long-term priorities and developing possible areas of need for further discussion and, ultimately, vetting and scoping. See Table 6.6 for what to look for in the practitioner phase.

Benefits at this stage include gaining significant traction in a single function (programme management, data engineering, data science and so on). Motivation comes from seeing progress from their efforts – especially on big, long-term projects that have been sitting on the to-do list for a while. Data teams that invest in deepening their functional skills are better suited to consult and advise stakeholders just learning about the data supply chain and their impact on it.

---

87 Apple defines this principle as follows: 'We believe that we need to own and control the primary technologies behind the products we make.' https://www.apple.com/careers/us/shared-values/brian-m.html

88 IKEA defines this principle as follows: 'We are not like other companies and we don't want to be. We like to question existing solutions, think in unconventional ways, experiment and dare to make mistakes – always for a good reason.' https://about.ikea.com/en/about-us/ikea-culture-and-values

Table 6.6 Best practices: practitioner

What is the new thing you will be trying?	By now, the mission and guiding principles have been used, and you are starting to adapt by learning what is working and what is not.
	Nothing *new* – this phase focuses on doing more of the same type of projects in the same functional areas. By deepening skills and continuing to enable the team to function well, you will include more accurate forecasting of data requirements within the data team and across the organisation.
What roles do the executive sponsor and the functional leads of the data team play?	The executive sponsor should play an increased role at this stage. They will help the leads scope new data projects and secure resources for a limited number of high-priority projects. It's good to involve sponsors from across the business to set the precedent for increased involvement when possible.
	The functional leads directly involved in delivering data services should be gaining increased comfort with data as an ongoing service scope and management. Extended stakeholders and even some data team members not directly involved in these projects have not yet been educated on or exposed to data as an ongoing service; however, they will also see the impact of the resources in action by hearing the stories of those involved.
Which culture shifts should you focus on?	✓ Establish mission and guiding principles for the team.
	✓ Forecast and prioritise data needs.
	☐ Develop core systems to ensure quality and consistency.
	☐ Encourage deep collaboration between sponsors and working groups.
	☐ Build strategic partnerships to support ongoing resource needs.
	✓ Adopt a 'resource-conscious', not 'resource-constrained', approach.
	✓ Constantly review ethics, compliance and standards of data strategy.
How will you know if you're ready to move to the next level?	Moving to the next level happens when at least three big data projects in a single function (programme management, data engineering, data science, etc.) have been completed successfully. There should be high confidence in the basic processes and systems used to deliver these projects. Those involved with direct delivery should feel confident enough to contribute to best practices.

Culture shift opportunity: forecast and *ruthlessly* prioritise

A solid data team is prepared to take on data service projects at level two. Now is the time to develop a running list of business unit-specific requirements spanning 6, 12 and 18 months in advance so that you can begin to prioritise your efforts.

After running a high volume of ad hoc projects that are impossible to scale, you know that investing in data as an ongoing service requires advanced planning to be successful. When choosing projects, use the Eisenhower Principle, picking those needs that are important and urgent.

Project management means being effective as well as efficient. Our time has to be spent on important things, not just the urgent ones. To do this and to minimise the stress of having too many tight deadlines, we need to understand this distinction:

In a 1954 speech to the Second Assembly of the World Council of Churches, former US President Dwight D. Eisenhower, quoting Dr J. Roscoe Miller, president of Northwestern University, said: 'I have two kinds of problems: the urgent and the important. The urgent are not important, and the important are never urgent.'[89] The 'Eisenhower Principle' is how he organised his workload and priorities.

- *Important* activities have an outcome that leads to us achieving our goals, whether these are professional or personal.
- *Urgent* activities demand immediate attention and are usually associated with achieving someone else's goals. They are often the ones we concentrate on, and they demand attention because the consequences of not dealing with them are immediate.

Knowing which activities are important and which are urgent can help us overcome the natural tendency to focus on unimportant urgent activities so that we can clear enough time to do what's essential for our success. This is how we move from the immature position of everything being 'Priority 0' into a more mature position where we can grow our businesses and careers.

As your ability to prioritise and plan matures, you're getting closer to establishing the roadmap for data as an ongoing service. This discipline helps you to anticipate needs and begin planning on a (somewhat) more efficient timeline.

Remember strategy-based needs identification
In Phase One: Scope the Project, you learned to identify data needs based on existing tasks or priorities. Using the three 'why' questions in Table 3.1, you confirmed that your selected need directly impacts your multiyear goals. The to-do list-based method for identifying early data projects is the bottom-up approach.

---

89 D.W. Eisenhower, 'Address at the Second Assembly of the World Council of Churches, Evanston, Illinois' (19 August 1954).
   https://www.presidency.ucsb.edu/documents/address-the-second-assembly-the-world-council-churches-evanston-illinois

The to-do list method is still an effective way to continue to identify data needs. However, as we continue to have a broader impact across the organisation and use an increased volume of data, viewing requirements using our multiyear goals as the starting point and drilling down from there to specific projects provides a more strategic view. The goal-based method for identifying data service projects is the top-down approach.

Table 6.7 demonstrates the how method and how it plays out as a data-as-a-service requirements list for your organisation. Table 6.8 helps you consider data requirements from the stakeholder's perspective.

The main benefit of strategy-based needs identification is that the link between projects and strategic goals is immediately clear. Because those strategic goals typically reach

Table 6.7 Two examples for using the 'three hows'

	Example One: Drive awareness of importance of developing data skills through the release of multiple case studies	Example Two: Reduce data usage costs through a data warehouse governance policy
1. What is the multiyear goal or strategic priority?	The strategic priority being addressed is to help all employees learn about the data supply chain and their role in it to better understand and support data as a service effort.	The strategic priority being addressed is to reduce data usage costs related to centralised data warehousing.
	(by)	(by)
2. How will you address this annual organisational goal in the next year?	Seeking to educate and influence executive functional leaders, provisioners and consumers of data in the organisation to map business questions to technical data requirements by framing case studies (targeted at key personas) with relevant job aids across the organisation.	Identifying the top users, process automation for cost monitoring, and determining ongoing monitoring and analysis processes.
	(by)	(by)

(Continued)

Table 6.7 (Continued)

3. How will the annual organisational goal be addressed through yearly department goals?	Educating senior leaders, provisioners and consumers of data about their role in the disclosure, transformation and consumption of data and the human and machine requirements necessary to curate data as a corporate asset through creating case studies and job aids.	Utilising the feature set of the data warehouse to set alerts for high data collection, partner with data scientists to analyse collected data and determine governance policies for ongoing data usage.
	(by)	(by)
4. How will you meet department goals through specific tasks, activities or needs?	Using a team of cross-functional stakeholders to review awareness and development materials for ease of use, consistency and effectiveness; amplify via existing marketing and HR/learning/ development efforts to drive awareness.	• Review data warehouse for good practices to support ongoing monitoring (i.e. setting alerts, determining thresholds, identifying efficiency opportunities). • Implement a standard report to run diagnostics. • Use governance policy to introduce a cost-sharing model among business stakeholders.

Table 6.8 Generating a data projects list

1. Consider possible sources of data requirements and begin generating a list of high-priority opportunities.	
To-do list-based data requirements	**Strategy-based data requirements**
↓	↓
↓	↓
↓	↓

2. Vet all potential data projects based on four criteria for success projects:
  1. The scope is clearly defined.
  2. The project is important *and* urgent.
  3. Resources identified in the business are ready and willing to participate.
  4. Education or training does not prohibit initial engagement or ongoing progress.	

(Continued)

Table 6.8 (Continued)

↓   ↓   ↓
3. Using your data requirements list, generate an engagement plan that organises needs by timing needs. Communicate this to the data team and sponsor group to begin the project scoping and securing the resource process. Revisit this at least quarterly to update.
Examples include: 1. Business analysts gather data to inform a strategy decision before a key product launch.   2. Data engineers must analyse usage in the data warehouse to manage unexpected cost spikes with long-term budget implications.   3. A data scientist must determine a clear product hierarchy across a suite of products and develop a process for accommodating new SKUs as they become available.   4. A programme manager must kick off a data strategy project and develop an ongoing process for monthly/quarterly requirements gathering and governance.   5. A programme manager must prepare awareness efforts (i.e. communications plan, stakeholder training, etc.) to drive stakeholder awareness, usage and compliance, educate new team members and drive awareness among extended stakeholders.   6. A governance project manager must manage data cataloguing efforts to support metadata and data steward activities.   *List all potential projects or data initiatives here.*   1.   2.   3.

farther than the next six months and are (likely) functionally interdependent, you can better plan what the data team might achieve for multiple business units and how the business units might augment resource gaps (via working groups, SMEs and so on). The exercise in Worksheet 14 will help you walk through this process.

The importance of leadership and prioritisation

The project list generated from Table 6.8 is subject to many common pitfalls, but a lack of leadership results in scraped knees for everyone. The executive sponsor must help change the perception that data projects are the sole responsibility of the data team, and they must perform a sponsorship role through their behaviours.

In this way, data projects are *business* projects, and there is joint accountability for their success. A strong, active, visible Primary Sponsor must support the project manager in building an engaged coalition. Does the executive sponsor help drive accountability among their cross-functional peers, from data migrations to platform changes to ongoing reporting needs? Do they help to hold working group members accountable? Do the senior sponsors know how to be good sponsors? Refer to Phase Two: Determine Resources for tips on managing sponsors and working groups.

To prioritise data projects based on what is important and urgent, determine an evaluation method to help facilitate conversations where everything is 'Priority 0'. Time to completion is an easy and uncomplicated way to sort projects, as illustrated in Table 6.9. However, many additional, important variables (that is, risk, cost, dependencies and so on) contribute to prioritisation, especially when trying to be ruthless. The data project lead and partnering groups must take the time to determine what those variables are and clearly define them. They may also want to assign weighting for those variables they wish to emphasise.

Ultimately, the number of variables discussed is what the culture (generally, leaders) demands or tolerates. Variables to consider using in the prioritisation exercise and tools to rank them are listed in Tables 6.7–6.9. Table 6.10 outlines potential considerations for evaluating data projects from a benefits perspective. Table 6.11 outlines possible considerations used to assess data projects from a risk perspective.

How receptive is the organisational culture to considering and using these variables to prioritise work? The answer to that question determines the operational pace, the depth of problem-solving efforts – not just data problems, but all problems – and the impact of data projects on the bottom line. A culture with a relatively low tolerance for deeper thinking will remain at a superficial level of impact. The trade-off between time and each conditional concern applied to your prioritisation method might be worth it, depending on the desired outcome. Sometimes, the fastest solution is (initially) the best one for some emergencies.

Those leaders willing to spend a little extra time considering business benefits and risks – even just a few of each, to start – will continue to uplevel their collaboration skills (because these discussions can be time-consuming and it can be hard to reach consensus), prove impact (as benefits will become more transparent and easier to prove) and mitigate risk (cross-functional thinking brings a deeper dimension, more inclusivity and additional creativity to problem-solving than succumbing to chronic fire drills within a single function).

Table 6.9 Example of prioritisation criteria for data projects, time

Open to-do list items	Risk: Medium
Projects with no set deadline for completion	Understanding of the data supply chain and strong leadership help mitigate stakeholder expectations. Frustration with unreasonable turnarounds for what seem to be simple requests or business requirements without context does not speak to the potential impact they may have. Are new report requirements truly a priority for the organisation? Not every data request drives a decision; should it be accommodated compared to the backlog of projects?

Remember not to take on additional data projects that are not a top need. All projects should have a clear and direct impact on business decision-making (where values are clear). |
| **Urgent items** | **Risk: High** |
| *Projects with deadlines of less than 6 months* | Even a good, well-scoped data project can miss a major deadline (approximately 50 per cent of projects do). If the project stalls, will it ultimately derail your ability to use and implement the final product?

List the data projects that must be completed in fewer than six months here.

1.

2.

3. |
| **Short-term priority items** | **Risk: Medium** |
| *Data projects with deadlines 6–9 months in advance* | Do not delay beginning the process of securing and scoping for projects on this list.

List the data projects that must be completed in six to nine months here.

1.

2.

3. |

(Continued)

Table 6.9 (Continued)

Long-term priority items	Risk: Low
Data projects with deadlines 9–12 months in advance	While the deadlines for these projects seem a way off, start thinking now about how you will align to and resource these opportunities. There might be related projects that will address adjacent issues in other functions. *List the data projects that must be completed in 9–12 months here.* 1. 2. 3.

Table 6.10 Benefit variables for prioritisation discussions

Benefits	(Proposed) Definition
Revenue enablement	Enabling increased revenue through data initiative that meets the business need
Customer partner experience (CPE)	Enabling increased c-sat through data that meets the business need
Return on investment	Realising greater ROI through data initiative that meets the business need
Productivity	Realised time savings on data corrections and system inefficiency because of improved data initiative
Cost avoidance	Reduced costs as a result of data initiative that meets the business need

You may choose to adjust the definitions and consider other variables. Once you determine your list of benefits, you must have a way of ranking them. Table 6.12 provides a continuum to consider.

Table 6.11 Risk variables for prioritisation discussions

Benefits	(Proposed) Definition
Level of investment	Measurement of costs required to implement a solution that meets business data requirements needs
Data quality risk	Degree of data quality risk due to the complexity of underlying factors
Breadth of cross-organisational impact	Number of business stakeholder groups and individuals impacted
Business readiness risk	Degree of business readiness and user adoption risk due to level of effort and degree of change management
Technical risk	Degree of technical complexity and/or resource skills and capacity constraints

Table 6.12 Proposed continuum for ranking benefit variables for prioritisation discussions

High = 9	High revenue increase	High CPE increase	High ROI	High productivity gain	High reduction in cost
Med = 3	Moderate revenue increase	Moderate CPE increase	Moderate ROI	Moderate productivity gain	Moderate reduction in cost
Low = 1	Minimal revenue increase	Minimal CPE increase	Minimal ROI	Minimal productivity gain	Minimal reduction in costs
None = 0	No increase in revenue	No CPE impact	No ROI impact	No productivity impact	No reduction in cost

You may choose to adjust the definitions and consider other variables. Once you determine your list of benefits, you must have a way of ranking them. Table 6.13 provides a continuum to consider. The risks relate to the benefits and should be considered in parallel.

Having dealt with 'priority 0' cultures, you can empathise with the length of time it may take to develop the prioritisation skill. From your personal experience, you might understand with greater clarity why some organisations shy away from it while others cannot live without it.

Table 6.13 Proposed continuum for ranking risk variables for prioritisation discussions

High level of investment (> $30M lifetime)	High level of data complexity	High number of stakeholder groups impacted	High level of business readiness risk	High level of technical risk
Moderate level of investment ($10–$30M lifetime)	Moderate level of data complexity	Moderate number of stakeholder groups impacted	Moderate level of business readiness risk	Moderate level of technical risk
Minimal level of investment ($0–$10M lifetime)	Minimal level of data complexity	Minimal number of stakeholder groups impacted	Minimal level of business readiness risk	Minimal level of technical risk
No investment required	Non-complex data	No groups impacted	No business readiness risk	No technical risk
Resource cost	Data hierarchy		Level of effort	Technology
Technology cost	Number of data entry points		Degree of business need	Capacity
Additional breakdowns	External data vs internal data		Level of change management	# of systems impacted

Prioritisation – like scoping, resourcing and managing work – is a skill that must be actively nurtured. It is one of the reasons why every organisation is on its own transformational journey, at its own pace – and why maturity curves are misleading.

Deepening a data team's *practice* at managing data projects is a commitment to emphasising effectiveness and efficiency. Depending on your organisational strategy and structure, these areas tend to be considered practices – much bigger than projects or processes – and they're often directly divided into additional sub-practices:

- IT operations.
- Service management.
- Data security.
- Data analytics.

- Governance.
- Leadership.

The phrase 'practice makes perfect' sums it up completely – the goal is to practise something regularly to improve. Since we're talking about data as an ongoing service, it makes perfect sense to develop our skills alongside that effort and practice in delivering data projects better every time.

The effectiveness of this kind of skill development is measured with key performance indicators such as quality, agility, compliance – and, ultimately, the ability to innovate. Every effort should be quantifiable to know your practice is working – that improvements are happening in that area.

Operational excellence is not just a corporate buzz term. When done well, it is a collection of intentional business behaviours in alignment with values such as organisational learning, making an impact and effective teaming (among many others). The drive to keep improving rests on two primary pillars: the systematic management of operations and the commitment to a positive culture that focuses on customers' needs and empowers everyone on the team. Empowered team members clearly understand goals and plans, feel secure in taking the initiative and come up with ways to fix problems.

Practising operational excellence requires you to see the concept of excellence as a culture that permeates everything you do rather than as an isolated initiative or event. Excellence, culture, quality – all of these – are outcomes of the system of processes and practices which you intentionally (there's that word again!) architect and live out day to day.

Those who have served on high-performing teams know the feeling of every member contributing at the top of their game and the satisfaction of doing great work. It's amazing when it happens.

LEVEL THREE: LEARNER

After practising with a few projects, it's time to bring the learnings towards integration efforts. As mentioned throughout this guide, data projects do not operate in isolation and must align with other parts of the organisation, as part of an ongoing service. After carefully scoping and ruthlessly prioritising projects, we've moved past ad hoc management. By changing the data team's practices before expecting others to get on board, we can continue to build on that work in addition to the processes already established by the early adopters and most enthusiastic stakeholder supporters. This is the stage when we will see the culture start to shift in the organisation in a quick and pronounced way.

Team morale will increase as data projects become valued across the organisation. Team members and stakeholders with early accomplishments can share their learnings and serve as data advocates to other staff members. Processes, platforms and systems developed by the data team might address different infrastructure needs in the organisation. Finally, sharing the goal of becoming data-driven will help unify people, improve ties to sponsors and provide cross-functional learning opportunities. See Table 6.14 for what to look for in the learner phase.

Table 6.14 Best practices: learner

What is the new thing you will be trying?	Having lived the mission and guiding principles thoroughly, you now have something solid to share with outside stakeholders. You are modelling behaviours and embracing commitments you want them to adopt.
	Bringing holistic data practices to other departments or functional areas. Solicit cross-functional input on priorities and decisions – especially when it involves competing commitments – even when challenging.
What roles do the executive sponsor and the functional leads of the data team play?	At this stage, the executive sponsor should function as an interdependent partner to their peers in the business when it comes to co-developing strategy.
	Similarly, data team leads in programme management and data science should align with and become deeply familiar with specific business units, actively consulting their primary business stakeholder on measuring and analysing strategy, considering deeper questions about their unique business problems and collaborating on potential solutions.
Which culture shifts should you focus on?	✓ Establish mission and guiding principles for the team.
	✓ Forecast and prioritise data needs.
	✓ Develop core systems to ensure quality and consistency.

(Continued)

Table 6.14 (Continued)

	✓ Encourage deep collaboration between sponsors and working groups. ☐ Build strategic partnerships to support ongoing resource needs. ✓ Think 'resource-conscious', not 'resource-constrained'. ✓ Constantly review ethics, compliance and standards of data strategy.
How will you know if you're ready to move to the next level?	When you've completed a series of projects with a partnering function, it's time to focus on a new skill set or functional area.

Table 6.15 Core leadership competencies for strong data projects

Business orientation (foresight) +	Process/Project management (oversight) +	Change management (insight) =	Strong data projects
Strategic thinking Holds the big picture view	**Computational thinking** Breaks down the big picture to execute	**Ambidextrous** Maps the many shifts needed to align strategy to execution	
• Defined business need • Focused on results, business benefits and outcomes • Manage and lead change • Forecasts, foresight • Understanding of role and participation in data supply chain • Stewardship	• Project definition and planning • Risk and scope management • Change control • Reporting and status • Assessment and feedback about project direction • Delegation and process management • Oversight: closing or ongoing improvement	• Behaviour-focused • Long-term vision • Adoption and Innovation-focused • Insight: ability to frameshift in observable behaviours and aspects of culture that need to shift • Successful cultivation of relationships • Impact analysis • Selling the change	

(Continued)

Table 6.15 (Continued)

	• Measurement and benefits realisation • Recognition, reward and retention of needed resources • Ongoing governance, disciplined decision-making	• Communication of mutual benefits and shared vision • Training and education of affected staff	

Which core competencies will be useful in running great data projects?

In Chapter 1, we reviewed human skills (for example, changing paradigms, computational thinking and dynamic learning) and the technical skills we need when operating the machine side of the supply chain (for example, cataloguing, hierarchy management, information processes and so on) as critical skills for effectiveness. Three primary measurable skills, abilities and behaviours are required to be a successful information management professional: a business orientation to employ effective foresight, project management to drive efficient oversight and change management to elicit the right insights. All data projects incorporate these competencies, and we gain these competencies through participation in all aspects of leading data projects.

Successful data projects result from combined effectiveness in three areas: business strategy and stewardship, project management in data functions and change management impacting culture and the pace of change (see Table 6.15). You might have strong competencies in each of these areas. For example, if you have eager, connected business stakeholders using your reports, you can more easily leverage their strengths in providing effective requirements to secure resources for combined reporting, deep-dive analysis or self-service reporting. Likewise, suppose you creatively utilise stakeholders and subject matter experts, such as managing adjacent workstreams or committees to deliver data projects. In that case, you likely have some core strengths for good change management. Use Worksheet 16 to assess competencies on your project.

No matter your strengths, you can typically use being scrappy as motivation to improve: find opportunities to uplevel the team with these competencies to strengthen your data project footprint. As you start to increase your influence across a function, a division or the whole company, think about the strong core competencies and the ones that still need to be developed.

Culture shift opportunity: develop core systems to ensure quality and consistency

A data-centric organisation runs on multiple concurrent data projects – varying in size, duration, structure and complexity. A repeatable system is needed to support these efforts. While some data projects likely relied on temporary solutions during the project's early days, system investments become more important when you start to scale. This is the time to standardise the process behind as many data efforts as possible. Table 6.16 summarises some common focus areas in developing support systems.

Table 6.16 Data supply chain: core systems

	Why is this important?	Key design factors
Requirements management factbook (Exhibits 6.1, 6.2)	• Promotes connection between high-priority business questions and data requirements (necessary context for data teams) and measures behaviour (necessary context for data scientists and analysts). • Addresses current data availability, addresses what business questions we cannot address and illustrates how we are addressing that gap. • Serves as a blueprint for early data stewardship conversations; creates a catalogue of top priority measures (necessary context for stewardship and governance and data storage estimates).	• Ability to keep track of business requirements, including data owner and early measure behaviour. • Ability to keep track of data stack development and self-service reporting status. • Tracks ongoing high-priority business questions that influence data strategy and technology investments.
Standardised training on data best practices (Table 6.17)	• Ensures all data team members and business stakeholders are equipped to use and interact with data tools. • Training data team members and business stakeholders as part of initial onboarding to data project work establishes the overall importance and commitment of the organisation to data initiatives.	• Stakeholders trained on key process points in addition to the culture of data-driven decision-making and organisational philosophy towards its application. • Individual roles and responsibilities were made clear, including calling out the collaboration of sponsors and SMEs.

(Continued)

Table 6.16 (Continued)

	Why is this important?	Key design factors
	• Data training should seek additional interventions through HR: infusing data-centred language in commitments, team and manager learning and development activities and so on, with relevant case study examples and behaviours that will drive the culture forward.	
Standardised recognition process (Table 6.18)	• Sponsors and stakeholders are contributing their skills and time. • Show groups they are valued and appreciated. • Demonstrate how their contribution impacts the overall outcomes.	• Say thank you in as many ways as you can – there are things you can give, say and do for team contributions. • Call out specific impacts of the work, not just the deliverables produced.
Reusable evaluation tools (Table 6.19)	• Consistent evaluation over time allows tracking overall improvements in data projects. • A period of reflection enables the ability to capture learnings and apply them to future efforts.	• Reflection should happen on a data project and annually (with sponsors and working members). • Reflection should encompass multiple perspectives – the team, stakeholders, sponsors, etc. • Reflection should cover both satisfaction with how the process went and its impact on the project's final deliverable.
Status, issue and risk management (sample risk register in Appendix E)	• Standardised and centralised project status helps everyone connect project tactics to the overall strategy, continually analyse for efficiencies and see around corners.	• Rigorous attention to identifying, defining and reporting against success measures. • Ongoing mitigation and management of risks. • Ongoing analysis and deep dives emphasise that no project is 'one-and-done'; they are all aligned to the big picture.

(Continued)

Table 6.16 (Continued)

	Why is this important?	Key design factors
Track and value data efforts (Exhibit 6.3)	• Track and report on the business impact (revenue, profit, sales, etc.) and project success measures (quality, time, cost, etc.) as an annual summary.	• Use industry-accepted project management variables to assess cost savings. • Report as part of quarterly and annual summary of contributions. • Be sure to communicate the value of specific business units to the business stakeholders and their respective sponsors.

1. Build a requirements management factbook

Most data teams and business stakeholders struggle to define and prioritise high-impact questions. When everything is important, nothing is important. Here are the things to keep in mind when designing a requirements factbook.

- *Define what business performance means for your organisation*: The term means different things to different data constituents. Performance management is typically composed of user experience, customer lifecycle management and performance management. Leaders determine the highest-priority business questions they need answered. From there, key performance indicators and other supporting measures are aligned to those questions to help prioritise what data needs to be managed in the centralised data stack. See Exhibit 6.1 for examples of common questions asked across the business that require ongoing measurement and impact strategic decision-making.

Exhibit 6.1 Example high-priority business questions

Business performance	Example business questions, decisions (*not exhaustive*)
User experience	• What is the level of engagement on each (product, service, etc.)? • How to structure the (product, service, etc.) to optimise engagement? • Does aligning product/service X with product/service Y matter? • What portfolio of online content maximises engagement? • What mix of features maximises engagement?
Customer lifecycle management	• What is the lifetime value of the customer? • What is the level of cross-engagement across (product, service, etc.)? • Most valuable points in (product, service) user pathways to cross-sell (product, service X)? • Most at-risk customers? Valuable customers? • What is the average revenue per unit?

(*Continued*)

Exhibit 6.1 (Continued)

Business performance	Example business questions, decisions (*not exhaustive*)
Performance management	• Performance (unique users, clicks, revenue, etc.) by product/ service, segment and geography? • What marketing campaigns have highest ROI? • What impact does the use of product/service X have on product/ service Y? • What is the wallet share by category and geography; how can we increase it?

- *Create a table of contents page that manages expectations on the availability of the data*: In Exhibit 6.2, most of the data managed by this data team is external (E) versus internal (I). The data team serves three primary data personas: data miners (data scientists and engineers), business analysts and managers. The team is in the initial stages of making data available. The project manager tracks availability in this table of contents to demonstrate transparency and quickly communicate how data is being prioritised and when it will be available (to whom).

Exhibit 6.2 Example table of contents for data availability

Business performance	Example business questions, decisions (*not exhaustive*)	Slide #	Available outside data stack	Enabled in data stack for... Data miners	Business analysts	Managers
User experience	• What is the level of engagement on each (product, service, etc.)?		E	✓	✓	FYxx
	• How to structure the (product, service, etc.) to optimise engagement?		E	FYxx	FYxx	FYxx
Customer lifecycle management	• What is the lifetime value of the customer?		E	FYxx	FYxx	FYxx
	• What is the level of engagement across (products, services, etc.)?		E	FYxx	FYxx	FYxx

(*Continued*)

Exhibit 6.2 (Continued)

				Enabled in data stack for...		
Business performance	Example business questions, decisions (*not exhaustive*)	Slide #	Available outside data stack	Data miners	Business analysts	Managers
Performance management	• Performance (unique users, clicks, revenue, etc.) by product/ service, segment and geography?	I	✓	✓	✓	
	• What marketing campaigns have the highest ROI?	E		FYxx	FYxx	FYxx

• *Create a page for each question*: Each business question will help tell some aspect of the data story that managers, data miners and analysts need to monitor and make decisions about the business. Managers (including executives, functional leaders and so on) look at overall descriptive and diagnostic data and ask questions about dips and spikes. They ask, 'How big is [what we are looking at], and what is happening?' Analysts look at the next level down to understand what is 'normal behaviour' for the measures, so they can provide an explanation when dips and spikes occur. They are asking, 'Why is this happening?' Data miners go even deeper to predict and prescribe the next steps.

The main elements on a page might include the business category, the question, two key data cuts for a specific time (and noting if the data isn't available), data source and the pull date. This context is crucial for effective data storytelling.

Different types of data (descriptive, diagnostic, predictive and prescriptive) show various decision choices and their impact on top-priority measures. Key performance indicators mark the destination, while additional measures (like geography, demographics) provide more context for informed decisions, similar to how a GPS suggests course changes in real-time. This is prescriptive analytics in action.

2. Provide standardised training on data best practices
Standardised training for the data team and business stakeholders on best practices and using data tools helps make data 'real' in the organisation. It also ensures that people are prepared to take advantage of any potential data projects.

What should the training cover? See Table 6.17 for a rough outline of what the data team and sponsors need to know to set themselves up for success. Review these suggestions against your current needs and customise. Most of the recommended topics are helpful for past data work and are good opportunities to partner with human resources and other functions. The right data best practices will also help your data team become better advocates and evangelisers of a data-forward organisation.

Table 6.17 General training about data

Topic	Key messages
What is being data-driven vs data-informed?	• Understand how data works and its overall value to the organisation, your division, and your team. • Share examples of what data is and is not (see 'Examples of being data informed' in Chapter 1). • Be sure to communicate the difference between ad hoc and intentional data requirements. Share specific case studies from your own data projects or those of peers. • Teach the five key ideas of making data projects work (from Chapter 2).
Driven by data – the concept	• Define 'Driven by data'. • Use Figure 6.1 at this phase's beginning to paint a clear picture for people of what a fully driven-by-data organisation looks like. • Why is this [project or initiative] worth doing? • What will success look like? What will you be able to do in the future that you can't do (well) now? • Highlight past experiences with data work; discuss return on investment, competitive advantage and the most constructive way of being 'scrappy'.
Introduction to scope	• Make clear that a well-defined scope is the foundation of a great data project. • Discuss how to identify data requirements – using the priorities outlined in the strategic plan and identifying project-based needs, outsourced roles and operational needs. • Review the strategic plan and other relevant departmental plans – and where to find that information. • Describe the four vetting criteria for data projects: (1) clearly defined scope; (2) important and urgent need; (3) that the education needed does not outweigh the benefits of the project results; (4) that sponsors and working members needed on the business side are ready and willing to participate.

(Continued)

Table 6.17 (Continued)

Topic	Key messages
	• Explain how to build a scope template.
	• Describe the ideal sponsor group and working member partnership in scoping data projects.
Introduction to securing resources	• Provide an overview of potential sources of data resources (see Phase Two: Determine Resources).
	• Describe potential models of project management– sponsor relationships.
	• Describe the sponsor and working member partnership in securing data project work.
Introduction to managing the work	• Provide an overview of the data supply chain and data lifecycle.
	• Give a phase-by-phase introduction to the key steps and checklist.
	• Ask and answer, 'What does it mean to understand data as a utility? And an asset?'
	• Review stakeholder motivations by phase. How can you help keep the team engaged through the dips and capitalise on the peaks in motivation?
	• Offer ideas for celebrating and recognising sponsors' and working members' contributions.
Fine-tuning ongoing change	• Call to action: describe the role of the sponsors in becoming data-driven.
	• Call to action: describe the role of working members in becoming data-driven.
	• Ask and answer, 'How will we communicate to ensure both parties continue to update each other on needs and project results?'
What does success look like?	• Ask, 'How will we know if we are a data-driven organisation?'
	• Reiterate what a data-driven organisation looks like.
	• Describe how to evaluate and measure the impact of the effort.

Use this training to prepare your organisation to create a stronger data culture:

- Provide training as part of onboarding for all incoming employees and interns.

- Make time in senior leader meetings to provide training to executives, paying specific attention to the role you expect them to play in scoping and securing resources for data work.

- Schedule annual or biannual all-staff discussions to share learnings from completed data work and provide space for dialogue about your overall data strategy. Use this to build buy-in to data-driven decision-making by highlighting data's impact on each department's ability to meet stated goals.

Design a standardised stakeholder onboarding process

The data team and business stakeholders are on board. They know how to use the data tools and drive the use of data in their teams. What about their extended stakeholders? Take the time to create standardised data training for extended stakeholders, which anyone can deliver. Doing so will save you time overall and reinforce that data is an official part of the employee's world.

A standard onboarding might include the data supply chain concept, what it means to translate business problems into technical requirements, the background and mission of the data organisation, the organisational chart and how teams work together, an overview of that year's strategic goal, how to submit data requirements, an issue ticket and so on.

3. Recognition process

Appreciating member contributions is perhaps the most important act as a project manager or an executive sponsor. Without proper thanks, the whole system is broken. A member who doesn't feel valued, appreciated or recognised will likely disengage or drop out altogether. Potential ways to motivate and recognise others were discussed in Chapter 5. As your efforts continue to scale, think of how you can make appreciation a stronger and

Table 6.18 Sample appreciation practice

Timing	Activities
Start of project	• Provide a letter or short video from the executive sponsor thanking them for their contribution.
	• Create a badge for the kick-off activity of the team. (Think of badges people can earn throughout the project.)
Ongoing during active project work	• Say 'thank you' at each meeting.
	• Find one way to connect the project to the overall ability to deliver on your mission and share it at the start of each meeting or call.

(Continued)

Table 6.18 (Continued)

Timing	Activities
Completion of project	• Provide a letter or video from the executive sponsor congratulating the team for their hard work. • Throw a celebratory gathering and invite the whole team. • Publish a newsletter article. • For resources sourced from the business, offer to talk with their manager, call out what they did exceptionally well and help them highlight the project's accomplishments. • Have sponsors sign an e-card for the working team.
Annually	• Offer recognition in year-end events and publications – acknowledge the impact on your ability to deliver on mission, ROI, project focus and team structure through your annual communications. • Invite consultants or third-party partners to events as VIPs. Offer complimentary tickets to each team member. • Use events to thank consultants and third-party partners. Make recognition part of the programme. As a sponsor, speak to the project's impact. Call each team member by name and role, and communicate their impact.

genuine skill in the team depending on the extent of the project and the contribution by the individual (see Table 6.18). Think of turning appreciation into an ongoing practice.

4. Design reusable project evaluation tools
Evaluation becomes critical as more data projects get added and your initiatives begin to scale. Take the time now to establish a transparent process to gauge the success of each project. As your investment of time in identifying and managing data projects increases, so should the desire to see clear and tangible evidence of its impact. As illustrated in Table 6.19, there are four evaluation categories.

A standardised process for evaluating all data projects is key to scaling efforts. This means preparing surveys, questionnaires or conversation guides that can be used across data projects with minor customisation. Smaller projects (that is, a fast report to address a one-off question) might not require a full-impact evaluation and should focus on categories one through three. Other undertakings (that is, developing a governance effort or a digital literacy curriculum and so on) should likely use evaluation in all four categories.

Again, much of this depends on your organisation's evaluation culture. Is it a key part of your strategy? If not, you may want to focus on establishing a process that effectively determines overall satisfaction with the process and the ability of the deliverable to meet the stated need and project scope. Both are fundamental indicators that the data project you are driving is worth your investment of time.

Table 6.19 Four evaluation categories

Satisfaction	Success of project		Vision impact
Satisfaction with the team interactions	*Meeting of stated need*	*Long-term solution*	*Increased capacity to deliver on stated goals*
Did those involved with the process feel satisfied with how the group interacted and the overall process?	Does the final deliverable address the challenge that it intended to resolve?	(Best conducted after 12–24 months). What impact has the deliverable had on the challenge it addressed?	Can you demonstrate an increased ability to deliver on your vision through the impact of the data project?
Evaluation in this category looks only at how the project went. At a minimum, all engagements should be evaluated on these criteria.	Evaluation in these categories looks at what a business unit (stakeholder) can do by integrating data into its strategy.		Evaluation in this category looks at the ability of the business unit (stakeholder) to deliver impact on its stated vision and goals as a result of the data project.

For additional insights into common areas of evaluation, see the section 'Evaluate and Acknowledge' in Phase Three: Manage the Work. Several methods for gathering feedback are described. Here are a few more approaches to consider:

- *Quantitative surveys*: Asking team members a standard set of questions establishes a good baseline and initial dataset to review year after year to see how satisfaction changes over time. Challenges include limited reach in terms of scope and diversity of answers provided. Additional comments can be superficial and hard to follow up on versus an in-person interview. Still, comments can be reviewed for high-level themes.

- *Qualitative surveys*: Questionnaire forms distributed to each participating member, from working member to external consultant to sponsor, can help generalise feedback categories. A standardised form can be used on all data projects, reducing the evaluation time required, although it will not allow for tailored questions to address project specifics. Challenges include a low response rate and the possibility of gathering only superficial commentary – and no opportunity to further probe into areas of interest. See Appendix E for example templates.

- *One-on-one interviews*: Schedule a dedicated time to sit down with each of your group members and ask specific, focused questions. This more intimate, conversational setting helps capture more information about specific areas of interest beyond what the project manager might feel compelled to report on through survey questions. Challenges include the possibility that members might 'sugarcoat' their responses because of the face-to-face design of this evaluation technique.

- *Group feedback or discussion*: An efficient way to gather a range of opinions and provide a sense of closure for the team, holding a single group feedback session can help foster effective dialogue between team members. Positioning this type of evaluation setting as an opportunity for learning and a time to determine how to improve future data projects is critical. Team members may be reluctant to provide candid feedback in some settings, so ensuring that members know you want to hear all feedback, good and bad, is very important. Challenges include the possibility that group members are not transparent in front of their peers or that groupthink takes over when members influence each other to present a unified perspective instead of individual points of view.

  - Design this conversation to run like a focus group.

  - Prepare five to seven open-ended, broad questions and a few prompts for each in case the group gets stuck.

  - Establish ground rules (for example, 'We want to hear all of the feedback – good and bad – and we encourage you to respond directly to each other'). Your role is to guide the conversation, but your primary goal should be observing and listening to the dialogue, not running it entirely.

- *Third-party interviews*: If you are ready to invest in understanding the success and impact of the project, consider bringing in a third party to assist you and conduct the evaluation process. While this approach requires an investment of time and resources, this technique enables an independent, unbiased perspective and often produces the most insightful, most constructive results.

  - One straightforward way to do this is to have a data lead distribute the survey(s) or conduct one-on-one conversations using the DCAM Maturity Model.[90]

  - On the other end of the spectrum, the data team might decide to invest in hiring a third-party consultant to help fully assess the team's (or organisation's) maturity. If you are asking a sponsor or other stakeholder to assist with this process, be sure to provide sufficient context on the project's goals, areas of interest for the evaluation and sample questions to guide their discussions.

5. Track and value your data efforts

Planning, measuring and reporting the value of data projects is critical to scaling data initiatives in your organisation. Remember to thank sponsors and stakeholders, and be specific about their contributions. Taking stock of what you value and appreciate and officially recognising expertise loaned to the cause is important.

How can you report the value and return on investment of data projects?

1. *Cost reduction*: Automation and the creation of an analytical ecosystem that is easy to use typically result in a significant decrease in the amount of time users spend on mundane tasks.

---

90 https://dgpo.org/wp-content/uploads/2016/06/EDMC_DCAM_-_WORKING_DRAFT_VERSION_0.7.pdf.

2. *Direct data monetisation*: Calculating ROI is typically a straightforward process that relies on revenue from selling data.

3. *Increasing revenue*: Data-driven decision-making supported by data analytics can be assessed regarding its impact on individual projects. For example, machine learning for customer segmentation may help focus advertising efforts on the most promising prospects, reducing marketing costs and increasing sales revenues.

4. *Faster time to market*: The user-time saving mentioned above can result in those users starting work on new initiatives sooner and delivering them earlier, which will have a quantifiable impact. For example, if a data scientist dedicates an average of five hours per week to prepare data and can reduce that time commitment to just one hour through new data and analytics initiatives, it would free up time for work on more complicated and innovative tasks. However, the time saved by these types of initiatives is not always limited to data teams; businesses may need to consider how everyone in the organisation will be affected to realise the full potential consequences of each project.

5. *Security*: This factor is calculated based on estimated losses that could occur if third-party access to critical business data is gained by others.

6. *Compliance*: This factor should be considered in terms of the potential loss to the business if it delays adoption or the initiative fails. A lack of investment in data initiatives or solutions can increase the likelihood of a data breach or fines for regulatory non-compliance (such as with GDPR). When a new compliance framework such as GDPR appears, there are usually few discussions on whether the company should comply. However, if putting it into numbers is necessary, we can use the same approach as with security.

How to calculate the cost of data

For simple, ad hoc projects, the easiest method to calculate the cost of data work is to break tasks down into their component parts and estimate the time and number of people impacted. Then, subtract this total from the time it takes when the task is automated. For example, a report took a week to assemble, data wrangle, prepare and analyse – you would summarise the time spent on each task versus automating some percentage of those tasks.

A before/after state of the operations of the report starts to become more apparent. You would end up with a formula looking something like this:

Automation savings = manual hours saved in 1 month * the total number of users * approximate hourly employee cost

Exhibit 6.3 illustrates a sample report process common among many analysts.

In preparation for a monthly business review, a programme manager must prepare data and analysis to communicate the overall health of an initiative they drive. The example above suggests that a *single* monthly report takes 291 minutes from start to finish. The effort is highly manual. The data comes in various formats. As the data comes together, it requires tedious checking and rechecking to ensure accuracy.

Exhibit 6.3 Sample report process: time and tasks

Time	Task
NA	Report template developed
NA	5 data sources identified
1 min	Reminder was sent to all data source owners to make data available.
	Collect data – in various formats.
15 mins	Data source 1: PDF format, requires manual entry of 7 rows
5 mins	Data source 2: TXT format, requires manual entry of 3 rows
40 mins	Data source 3: XLS format, requires manual entry of 20 cells
10 mins	Data source 4: PDF format, requires manual entry of 5 cells
40 mins	Data source 5: XLS format, requires manual entry of 7 rows from 15 individual tabs
60 mins	Quality check, initial analysis
60 mins	Group quality check, draft analysis for team
10 mins	Analysis prepped and integrated into an appendix for monthly business review with caveats.
20 mins	Issues/Errors report to data sources for discrepancies.
30 mins	Ticket logged with the data team for additional data needs for the following month.
291 mins	**Total time consumed by manual reporting activities**

As with most repetitive data efforts, this data is likely put into a spreadsheet that automates the quarterly and annual averages and totals the programme manager needs to gauge overall health. While initially helpful, attempts to automate complex information often fail. They require constant maintenance and frequently break if more than one person attempts to use the file. That time is not factored in, but these breakages happen – and over time, the time spent on them must also be factored into the report task list.

While this is just a look at a single report, these kinds of manual, decentralised data efforts can take a toll on individuals, teams and organisations. Detailed task analysis for *all* data activities can give valuable insight into all disclosure, transformation and consumption efforts.

Once the report is delivered, data plays a very active role in increasing value in the business (increased sales, improved performance, reduced returns, enhanced customer satisfaction and corporate reputation). It is essential to consider all possible scenarios in which data-related projects (such as data science projects) could benefit and lead to meaningful business impact.

Culture shift opportunity: encourage deep collaboration between sponsors and working groups

To integrate data successfully and intentionally as a service to all parts of the organisation, sponsors and working group members must be activated, engaged and supportive of creating a data-first culture. This section will review key principles for engaging leaders and resources in this culture shift (see Figure 6.3) and describe the necessary systems to support a mature and thriving data culture.

Figure 6.3 Sponsors and working group members in data-driven model

Activate sponsors

Successful, organisation-wide data initiatives are best served by sponsors in areas critical to creating cultural change through data or digital upskilling: human resources, finance and marketing. As mentioned, sponsors play a crucial role in change, but information alone will not make them successful. Project managers must enable sponsors by clarifying the sponsor role, sharing supporting research and data, offering coaching and encouragement and helping with tactical responsibilities to set up the sponsor for success. By facilitating specific sponsorship activities and action steps, change practitioners become vital partners in driving successful change outcomes.[91]

---

[91] https://www.prosci.com/resources/articles/primary-sponsors-role-and-importance

In the data-driven model, sponsors serve two primary roles:

1. *Scoping projects*: With diverse functional backgrounds, sponsors can likely quickly and accurately scope many of the data engagements in partnership with working group members.

2. *Determining resources*: Most sponsors help to 'raise resources' along with other kinds of support when kicking off large initiatives. Integrating data strategy, effective stewardship and ongoing governance is part of a business stakeholder's core responsibilities. Ask them to serve as connectors, providing the introductions and network required to drive data culture from the inside (for example, individuals, intermediaries, teams and functions) and the outside (third-party partners, professional organisations and universities).

Table 6.20 highlights four principles for engaging sponsors.

Getting the most out of sponsor involvement

Senior leaders manage profits and losses but also have 'personal balance sheets'. They have connections. To break down the silos that naturally form when we move quickly to get things done, leaders (at all levels) must rely on their networks to invest in important initiatives of the organisation they support.

Tapping into the sponsor group's potential assumes they have been briefed and trained on the basics (see the earlier section on training sponsors and working group for a suggested outline of topics to cover). Here are a few tips for how to get the most out of the sponsor group's involvement:

1. Add data-specific behaviours to recruiting copy and job descriptions for sponsors, working members and SMEs. Seeing these behaviours modelled across all functions at all levels does much to help evangelise the needed culture shift.

2. When new sponsors join, ask them about their ability and willingness to determine resources for your project.

3. Propose a small percentage of the sponsor's budget as part of their participation goal. Suppose they pay into the common architecture, which feeds a self-service reporting layer they depend upon for decision-making. In that case, they have a vested interest in understanding the data supply chain and mapping business problems to technical requirements – their interest and support shifts from passing to highly engaged.

4. Add specific annual assessment language to sponsor members' reviews.

5. At each status meeting, recognise what data budget was raised and what specific business unit resources were added to the initiative's efforts.

6. The project website and the annual newsletter acknowledge data integration contributions, referencing individual sponsor members involved in securing resources and budgets.

7. Phase One: Scope the Project proposed that all data projects should be high-impact, urgent and closely aligned with organisational priorities. Do not ask for introductions or leverage a sponsor's network for low-priority, low-impact work.

Table 6.20 Four principles of sponsor involvement

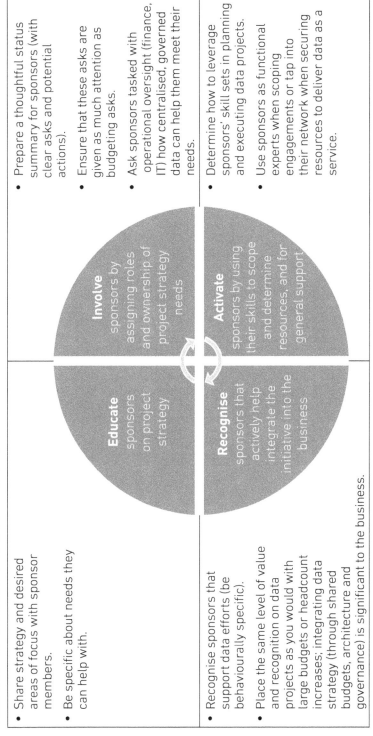

- Share strategy and desired areas of focus with sponsor members.
- Be specific about needs they can help with.

- Prepare a thoughtful status summary for sponsors (with clear asks and potential actions).
- Ensure that these asks are given as much attention as budgeting asks.
- Ask sponsors tasked with operational oversight (finance, IT) how centralised, governed data can help them meet their needs.

- Recognise sponsors that support data efforts (be behaviourally specific).
- Place the same level of value and recognition on data projects as you would with large budgets or headcount increases; integrating data strategy (through shared budgets, architecture and governance) is significant to the business.

- Determine how to leverage sponsors' skill sets in planning and executing data projects.
- Use sponsors as functional experts when scoping engagements or tap into their network when securing resources to deliver data as a service.

The circle diagram contains: **Educate** sponsors on project strategy; **Involve** sponsors by assigning roles and ownership of project strategy needs; **Recognise** sponsors that actively help integrate the initiative into the business; **Activate** sponsors by using their skills to scope and determine resources, and for general support.

Doing so may risk the appearance of ingratitude in relation to the sponsor's support and connections.

8. As part of the annual planning process for operational goals, engage sponsors with functional expertise to identify data solutions to the challenges being considered. Look at project-based solutions, working solutions and outsourcing opportunities.

9. Challenge sponsors to find appropriate opportunities in the next iteration of the strategic plan to integrate data resource generation as part of the overall strategy for each goal.

These tips can turn your sponsor group into a team of data champions. When paired with the right working group members, the sponsor can also be an important asset in scoping project work. In a ProSci study of sponsorship, 73 per cent of respondents with extremely effective sponsors met or exceeded objectives compared to 29 per cent with extremely ineffective sponsors.[92] Effective working and sponsor groups make the ideal dream team for data initiative scoping. But both teams need help to get the most out of their expertise. Get more out of sponsors by:

1. *Defining the ideal outcome*: Train sponsors to expect status against the scope statement. Guiding them towards the core components of the scope – what will and will not be accomplished, what success looks like (see Phase One: Scope the Project for more details) – establishes guardrails for the conversation, prevents scope creep and keeps everyone on track.

2. *Reminding them of the most common success factors for data projects*: These include having a clearly defined scope, picking important and urgent projects, validating that you have the knowledge needed on the team and ensuring that the working group and sponsors have adequate bandwidth to be successful on the project. See Phase One: Scope the Project for a refresher on the criteria for success.

3. *Focus on communicating needs and not the solution*: Help sponsors accurately scope a data project by describing what success looks like. Don't simply list the items or things you hope will be produced. Instead, communicate what they will be able to do because of this work. Frame the pain – describe the pain being felt (*remember what it took just to get a manual report completed?!*) and not the medicine needed. The sponsors can be very effective in helping to craft the right solution together. This also ensures that they feel ownership of the selected problem-solving approach. Remind them of any unique factors of the organisation that might impact how the project is designed (for example, working with specific demographics, legal constraints, budget constraints and so on).

   ○ *Don't lead with*: 'We need a self-service reporting layer.'

   ○ *Do frame the issue*: 'The current data architecture prohibits business units' uniform access to metrics, forcing analysts to set up their own data delivery mechanisms and manually assemble reports. This contributes to quality and accountability issues throughout the data supply chain.'

---

92 Karen Ball, Tim Creasey, Kent Ganvik, Emily Hazelton, Lisa Kempton and Robert Stise (eds), *ProSci Best Practices In Change Management*, 11th edition (ProSci 2020).

4. *Give them space to apply their expertise*: They are serving as sponsors because they are experts in their fields. Work with them to map out the key information, then rely on them to put it in and lend their experience.

5. *Focus on staying specific*: For each item included in the scope, push to explain what needs to be done, to what extent and over what period. For example, don't just list how many discovery interviews; also include the initial findings and recommendations for next steps.

Activate the working group

In a data-driven organisation, the data team and working group will serve three main roles:

- Scope data projects (with the partnership of sponsors) to meet the needs and strategic objectives of the organisation.

- Secure support for data projects with or without the support of the sponsors.

- Manage data projects – through acting as day-to-day contact, technical specialist or implementation team – that impact their team or department.

Creating a data culture from staff to organisation

There is a whole other book to be written on this topic. Here, we will only cover a small part of what you can do for your data team *today*.

A lot of the energy and attitude of the culture is taken from the top. In Exhibit 6.4, how the sponsor 'shows up' sets a tone for the rest of the team and the organisation. Their presence emphasises that the mission is important, how it matters to other groups (or not), what's coming next. A sponsor can make or break an initiative.

Start small by modelling the change you want to see. Prove change in your own team first. Use these techniques in your team to build interest in and commitment to data-first behaviours. Then, work with stakeholders to adopt these practices.

- *Acknowledge what can be accomplished by leading with data*: Establish that those group members who successfully use data to tackle their to-do lists will be recognised for using strategic data to drive change. Place extra emphasis on projects that have defined and achieved success-relevant measures.[93]

- *Make enhancing data culture, modelling data-driven decision-making and integrating data in every business strategy an explicit part of individual employee goals*: Outline the specific use of data support for that individual (data team member, workgroup member, sponsor, partner and so on) during goal setting (see Table 6.8) and align goals with the data needs which the sponsors have established and projected for 6–12 months out. Encourage adding stretch goals that can be attained through the use of data.

---

[93] The use of *relevant* is critical. There are times when people claim victory over the wrong measures, which will ultimately lead to underperformance of the individual, team and overall strategy. Find examples of people using data effectively and highlight their work as exemplary.

Exhibit 6.4 Case studies of sponsor involvement

Effective sponsorship:

'My sponsor was the CIO. She was involved throughout the entire project and very transparent, present, engaged and committed to participating as needed to ensure ongoing success. She demonstrated change behaviours in sight of impacted groups. She inquired into whether they were finding the change difficult and how they were overcoming that difficulty. She was present to tell employees how they will be supported if they are struggling with the change.'

Poor sponsorship:

'My sponsor was the CDO. Due to an aggressive travel schedule supporting sales, he had minimal knowledge to make decisions. Continual deflection and prolonged decision-making resulted in virtually no decisions. Empty promises during a project caused fall-out, turnover and poor culture. When he did engage, he would micromanage where he thought he could make an impact. Constant requests for detailed spreadsheets with tight turnarounds brought productivity to a standstill. Solid recommendations were made by the team and replaced by these spreadsheet exercises, which decreased morale. By focusing on some of the most remedial tasks, he caused misalignment of the project's communication/purpose/mission. He was ineffective as a sponsor and a leader, making his position extremely difficult to work with.'

- *Give them the tools they need to be successful*: People need training, tools, inspiration and motivation to play their part in making a data-first culture. Give them what they need to be successful.

- *Training*: Provide standardised training on data best practices as a mandatory part of team and/or organisational onboarding. The data revolution requires an army. Everyone on the data team, every stakeholder from the working group and every sponsor needs to know how to identify their own data needs.

- *Tools*: Keep a library of any known data resources, whether virtual or hard copy, roles and responsibilities forms, process overviews, meeting agendas and links to publications about the impact of data (DAMA and EDM are popular data standards organisations, see Professional Resources for overviews). This keeps easy-to-use tools readily available.

- *Inspiration*: Schedule annual or biannual meetings for all members during which you can discuss your team and organisation's use of data. Use this to share lessons learned, best practices and fun contests and to promote peer-to-peer dialogue.

- *Motivation*: Build excitement around the use of data. Send out an all-member email when a new resource has been secured. Recognise those members who have used data successfully with announcements at staff events or through individual performance reviews.

- *Accountability*: Define what success looks like in this area and acknowledge successes and failures as a team. Learn together.

LEVEL FOUR: SKILLED

At level four, you are a skilled user of data. The data team, working group members and other resources across functions can apply data to other areas of need. The movement is spreading! You are well versed in running great data projects and have used projects to address needs in multiple parts of the organisation. You are ready to try more complex data services. See Table 6.21 for best practices.

What benefits are you seeing at this stage? As would be the case in a mature data-forward organisation, your team and sponsors will continue to partner. You'll build and strengthen more external relationships or partnerships as you bring more and more resources into the data movement.

Table 6.21 Best practices: skilled

What is the new thing you will be trying?	Continuing to adopt and adapt guiding principles additive to your current organisational culture moves you towards becoming more data-forward.
	Add a new data service model, such as an experimentation layer, or consider bringing in a new business unit stakeholder.
	This is a good time to reach out to a human resource business partner to understand how to infuse organisational competencies, learning and development, onboarding experiences and recruiting efforts to see how your best practices with data might influence the organisation as a whole, not just your stakeholders.
What roles do the executive sponsor and the functional leads of the data team play?	The sponsors and the working group should be comfortable working together in all phases of data use – scoping the project, determining resources, managing the work and fine-tuning the ongoing change. Not all data projects will require joint oversight at all phases, but with cross-functional, organisation-wide projects, you will want them to work hand in hand.

(Continued)

Table 6.21 (Continued)

	The executive data sponsor should start to reach out to their human resource peer. With their skills and enterprise-wide impact, the human resource leader is an ideal partner in connecting data, technology and people and cultivating collaboration.
Which culture shifts should you focus on?	✓ Establish mission and guiding principles for the team.
	✓ Forecast and prioritise data needs.
	✓ Develop core systems to ensure quality and consistency.
	✓ Encourage deep collaboration between sponsors and working groups.
	✓ Build strategic partnerships to support ongoing resource needs.
	✓ Adopt a 'resource-conscious' approach, not 'resource-constrained'.
	✓ Constantly review ethics, compliance and standards of data strategy.

Culture shift opportunity: build strategic partnerships to support your data service

Why partner? The most important thing you can get out of the right partnership is a steady pipeline of high-quality talent from a sponsor-stakeholder who understands what you want to accomplish and how. Building a strategic partnership with human resources can amplify data's impact on and in the organisation and connect several strategies to a single, shared organisational goal – becoming increasingly data-forward.

What does a strategic partnership look like?
Every successful strategic alliance will share a number of common characteristics, as shown in Table 6.22.

Culture shift opportunity: think resource-conscious, not resource-constrained

At this stage, data has fully integrated into the strategies of most businesses. You are starting a partnership with human resources to influence adding data-forward language to existing leadership principles, organisational competencies, performance goals, manager training and recruiting practices. While you'll see many benefits from having different people in different departments and roles involved in the data projects you are driving, you'll want to consider whether it's time to identify owners for more permanent pillars of the data initiative: data science more formally aligned to the business needs, strategy, governance and a project management office.

Table 6.22 Characteristics of long-term strategic partnerships

Consistent communications	The most critical element of communication in a strategic partnership is recurring, pre-set check-ins – at least once a quarter. • Create space to discuss upcoming needs. • Share feedback from past events. • Serve as a sounding board for general knowledge sharing. • Make sure these happen regularly – partners must see that they are the go-to resource to help you meet your goals.
Mutually beneficial activities	Recognising what each party is getting out of the relationship is key to creating value. • You are getting the assistance that you need. • Your strategic partner is strengthening their data culture and community, doing more with limited resources and providing their group members with robust development opportunities and career paths.
Support reaching all parts of the organisation	The areas of support – financial, general allocation of aligned resources and sponsorship – should reach multiple or all parts of the organisation. Using this support for programmes and operations is a key part of the structure of a successful partnership. • Allows you to expose your partner to all parts of your operations. • Provides multiple types of data projects; various activities along the data supply chain ensure stakeholders and engagement across multiple functions. There is, literally, 'something for everyone'. • Having many staff within your organisation trained and ready to use data – a key component of becoming data-forward – means more data stakeholders can become engaged and receive higher support. No management bottleneck is created by all opportunities being channelled through a single project manager.

(Continued)

Table 6.22 (Continued)

Financial, hands-on data skilling and data as an ongoing service	Partners should be supporting the data operations by providing a full spectrum of support – financial, resource alignment, hands-on learning and general skills-based training – by using and leveraging ongoing data services.

In Phase Three: Manage the Work, we built processes and systems to bring data organisation-wide. If you opted out of that, now is the time to consider those benefits. If you already have a robust stakeholder programme with the business, is now the time to partner more closely with human resources? Expanding data's influence means more resources. Consider your request for additional data team members, potential stakeholders and sponsors. Remember, data is the gift that keeps on taking. It's important to remind new members and sponsors of the cost of maintaining data as a utility and conserving it like a natural resource, and how it must be cultivated as an asset. Data must be constantly curated to remain relevant, timely and clean.

The main benefits of centralising data stack management, data strategy and governance efforts in one team are:

- *Consistency*: Incoming requests and requirements from the business will come in with similar scope information. Project managers on the data team will have a common approach to managing the work.

- *Efficiency*: Having lived the mission, evolved guiding principles and determined a method through your first several data projects, there is no reinventing the wheel, constantly starting from scratch and instant overwhelm – no more bumbling through. You have systems and practices that work and can continue to improve and scale as you serve more stakeholders.

- *Leverage*: Thinking creatively and holistically about every relationship and every project means you can extract more value with less effort.

The drawbacks to centralising are generally perceived as a loss by business stakeholders with their own data efforts. Business stakeholders building their own data silos might feel less motivated and ownership of the data movement you are building because they are no longer as autonomous. However, everyone wins if they get the right tools, support and voice at the table to prioritise their needs (all met with timely and transparent communication) from the centralised effort.

Be resourceful, not cheap

With a few projects completed, you have a track record. You are known to invest when it counts. Your sponsor group, working group and broader organisational network are ready to help because the value you provide in return is clear. Through your approach to leading data projects, you have demonstrated the difference between resourceful and cheap.

Cheap is when you are chasing every project, taking on every ask and delivering a single solution – once. Resourceful is taking a broader view of what is being asked and finding a way to provide ongoing value. Cheap is when you duplicate work and continue to execute highly manual processes. Resourceful is finding a way to automate a portion of every solution you provide, iterate efficiency with each version, ruthlessly prioritise changes and (most importantly) communicate the value of what you provide in ways that impact the bottom line.

When you're resourceful, sponsors, stakeholders and peers see how smartly you apply available resources. You are begging for budget and horse-trading for headcount. You are strategically leveraging your network to make win–win impacts on the organisation. This is how you approach all your data initiatives – with careful consideration and appropriate investigation.

KEY POINTS

- Transformational work is about managing ongoing change – like tuning into the right frequency. Managing continuous change requires support from the whole organisation to work in unison to make each effort successful.

- Becoming data-driven through data as a service requires a serious investment of resources – financial, staff, equipment, services and so on – and scaling your efforts will only increase those topline demands.

- Cultural shifts can take years. That is why diagnostic and maturity curves depicting each phase as equal (as mentioned in Chapter 1) are incredibly misleading.

- An organisation could be at level two or three for quite some time, and happily so – the pace of change depends on the business pressures to move from one level to another and healthy internal mechanisms to help enable that shift. This book has outlined a strategic approach to tuning into the right frequency and explains when a cultural shift might be considered.

- Committing to data as a service means committing to ongoing organisational transformation. Increasing data interdependency with business strategy means constantly aligning your organisational structure, talent, leadership and culture with your business strategy to sustain change and keep pace over time. That is why seeing a path through the forest and knowing where you can expect to be as you develop is important.

- Common external pressures and internal requirements to meet the changing demands that spark change require ongoing alignment – much like tuning into a signal frequency.

- Incremental steps help highlight one notable change at a time, which will help preserve quality and maintain forward progress. They are: observing (through exploration and gaining understanding); practising (learning by doing); learning (expanding efforts and iteratively improving); and becoming an expert (by leveraging insights across disciplines).

CONCLUSION

The *Driving Data Projects* guide has laid out many opportunities and challenges you'll face when kicking off a data project as the first step in the data revolution. Anyone can complete a project and delight a stakeholder. To seed a movement for change, to shift mindsets and culture, requires a method to become intentional with the change you want to make and constantly follow through.

This chapter reviews your role in the data supply chain and what not to forget. It highlights areas of interest and helps you reflect on choices of particular importance. Take the time to remember why seeding a data revolution is vital for your data team. Document the goals for your data project as part of a larger strategy.

YOUR ROLE IN THE DATA SUPPLY CHAIN

Having worked your way through this guide, you have learned:

- That data is information and philosophy. It prompts emotions, requires a unique language, transforms into art and (most importantly) must travel along the supply chain, undergoing many transformations, before it can be called an asset.

- That there is a human and a machine side to the data supply chain, and we must become more accountable as data teams and stakeholders for how data is acquired, managed, used and disposed of.

- How to scope a problem into an opportunity by aligning your project to multiyear goals and how to vet it against criteria for successful projects.

- New ways to think about resourcing projects through data teams and working groups, the importance of sponsorship and several ways to align with and leverage team and group members most effectively.

- Strategies and tools to get everyone on the same page, ensure the project stays on track and amplify clear human–machine contributions at each delivery phase.

- Strategies and tools for fine-tuning ongoing change in a way that increases your capacity to do more and amplify your data culture as part of your project delivery.

- How to play and adapt by challenging yourself to new ideas through reflections, questions, worksheets, checklists, case studies and tools designed to help you and your team brainstorm, build, problem-solve, plan and make decisions about each data project.

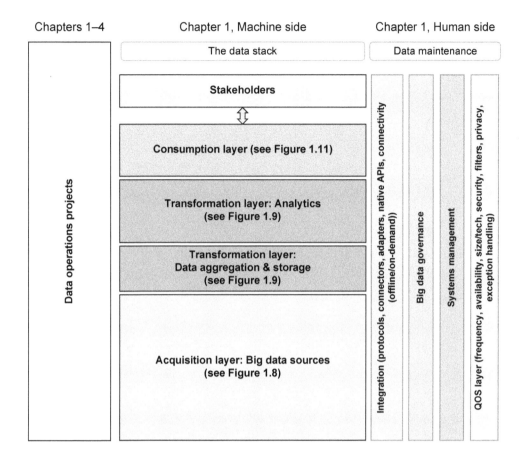

Chapters 1–4   **Chapter 1, Machine side**   **Chapter 1, Human side**

REASONS TO DEVELOP DATA PROJECTS STRATEGICALLY

Data projects seed the revolution of change

Taking time to think about how the data projects you drive could shift your organisation's culture is crucial to understanding what mindset shifts need to happen, the specific upskilling needs of impacted groups and the business value and benefits communications needed to create sustainable change. A well-considered approach for your data projects lowers the risk of missing key steps and provides opportunities for strategic alliances to deepen integration efforts.

The data team combines several disciplines – data science, data engineering, technical programme/product management and change management. Before tapping into your stakeholders' needs, learning the motivations of each area of the data team – as we learned in Phase Three: Manage the Work – is important to tap into the desire for change, and it is often overlooked.

Effective change starts at home. Having the data team operate with a singular mindset for driving change helps stakeholders hear (and learn) a clearer message.

Communicating the data supply chain to reflect the desires *and responsibilities* of internal and external stakeholders helps everyone look at data from a new perspective. While what needs to happen might be obvious to you, every other function will have another opinion until everyone can understand and agree on what needs to be done. From there, the project lead motivates others to do their part by making the changes necessary to nurture data as a corporate asset (not more static in the system).

Everything starts and ends with data. Therefore, every function must start to think of itself as structured around it – a radical shift from how business has operated in the past. Such a mindset shift is how we move from riding a bicycle to driving a Formula One car and catching up to the speed of business.

Data brings everyone together

Driving your data projects strategically creates opportunities for collaboration around driving impact. More importantly, ongoing collaboration and strategic alignments are important silo-busting behaviours that can result in co-goaling and joint long-term budget investment.

Generating a to-do list to identify needs bottom-up and using multiyear organisational goals as a starting point aligns data work top-down to maximise impact. This sounds straightforward, but it is tough to pull off. Why? Because most people's views on how data will transform their business differ. Neither the project lead nor the data leader start at the same level of understanding of what it means to steward the data supply chain into a corporate asset. Some stakeholders think data teams can sit in the corner somewhere, bolted on to the organisation, and still generate value. Others think the entire concept of data is too 'black box' for them to understand.

To get the most from the data team, data (that is, data science, business intelligence, insights, analytics and so on – however it is referred to in your organisation) must be regarded as a dominant discipline. This is why human resources makes an ideal strategic partner and why the mechanisms of human resources (for example, values, performance measures, leadership training and so on) make ideal areas for infusing data-centric language and competencies. Aligning the data team's impact around this perception and determining the ideal approach as part of the data project strategy will help position you for success.

Data lays the foundation for implementation

By developing a project-change method to inform your approach and priorities, you create foundational plans and practices to execute your data project. The rigour and inclusiveness of your approach will lead the business in the right direction. It will also provide a benchmark to reference when new opportunities and setbacks appear, requiring a shift in mental models, ambidexterity and the ability to adapt quickly.

Evolving your data supply chain in line with your understanding of the technical architecture serving current organisational needs is crucial to scoping data as a service (Plan in Phase Three: Manage the Work). Getting an early draft requires a lot of detailed thinking, cross-functional collaboration and setting aside egos around a sense of ownership. Thinking through current and future states requires thinking step-by-step through execution in both data- and machine-driven setups across the organisation (or whatever your scope may be). You are thinking through how the data will flow through the 'pipes'.

The strategy for your project will adapt over time as technology and business needs shift. Regardless of your approach and the practices you have set up, your message provides a solid foundation to kick off additional data projects and fine-tune ongoing scaling efforts.

Principles drive outcomes, early

Organisations tend to push out investment in data teams until end-to-end investment becomes expensive and unavoidable. Business stakeholders know they cannot operate without data, making earlier investments necessary. To mitigate the overwhelm, establish the data team's mission and priorities early (Observe in Phase Four: Tune the Change). These tools might seem like a waste of time, but they come in handy when the team encounters the inevitable resistance and challenges that accompany change.

Typical priorities to consider early in the project are as follows:

- Observe and document current state.
- What needs to change? Why? Why now?
- What are stakeholders' expectations? Are they aligned?
- How will these expectations be realised?

Data brings the objective into perspective

Using mission and principles to frame objectives helps drive cross-functional perspectives. Business priorities drive many *can-we* questions. Legal (*should-we*) and IT/engineering (*how-should-we*) drive privacy, security, trust and other ethical concerns about the data asset being considered.

When deliberating on the context of the change that needs to occur regarding a data project, think through current dependencies in your data supply chain. Table C.1 illustrates a few example data supply chain dependencies with real-world examples.

Data projects drive the revolution's message

Data projects are a lot of work when done strategically, aligned and fully embraced by all stakeholders. Getting to that point requires four critical components: well-defined success, active and visible sponsorship, and a well-blended project-change

Table C.1 Data supply chain dependencies with examples

Data dependencies	Examples
Usage limitations from data not fully owned.	Restaurants might need to licence third-party datasets to comprehensively view foot traffic volumes, nearby crime history, real estate development and other data sources, to inform their lease negotiations.
Digitising all aspects of the business to become data-driven end to end.	Hospitals must digitise patient records, and face interoperability, cybersecurity and privacy issues.
New roles, competencies and skills are needed across the whole organisation.	Employees and managers must gain digital, analytics and organisational transformation skills to become more effective in their roles.
New laws and regulations that previously did not apply.[94]	The European Union's General Data Protection Regulation (GDPR) says that individuals own their personal information and thus presumptively have the legal right to control it, and who can use it is a matter for them to decide. The specific rights that apply depend on the data type, especially highly sensitive data. As more US states adopt this standard, the implications of this fundamental shift in the underlying philosophical framework regarding data privacy protection will be profound in the years and decades to come.

management approach. The communication and training supporting these elements provide the foundation for a solid communication platform.

Sponsor and group-member needs help drive the communication plan for the project phases, which can be tailored for newly impacted or extended stakeholders, who will have different priorities and (perhaps) new concerns and mindset shifts they must now adopt.

Converting your project strategy into more consumable content such as onboarding training, FAQs, newsletters and even informal videos by the executive sponsor can all help educate and upskill the group as it continues to broaden – this is when you know your revolution has momentum!

---

[94] The United States has historically allowed businesses and institutions to collect personal information without express consent, while regulating those uses to prevent or mitigate harms in specific sectors.

Data amplifies decisions: good and bad

Any data project is about choices. Should data be centralised, federated or decentralised? How much of the reporting should be automated? Which measures should be governed versus ungoverned? Is it worth outsourcing our data warehousing? Should we build our own metadata solution? You have not accomplished anything meaningful if you try to execute everything by avoiding trade-offs.

The scope you set, the priorities you commit to and the trade-offs you make become the guide you follow during the challenges of driving a data revolution. Correct choices are critical initially because they help you gain credibility and momentum. Wrong decisions predictably impede the potential for future investment as well as progress in your career.

Make more correct choices by:

- Taking the time upfront to plan. (See Phase Three: Manage the Work, 0. Plan.)
- Investing time in educating the main sponsors and group members about the data supply chain and their responsibility in it. (See Chapter 1, Data Foundations, The machine side.)
- Seeking both internal and external experts to gain a broader perspective on the problem you are trying to solve. (See Phase Three: Manage the Work, 2. Discovery.)
- Making the data team your advisory group in addition to a key stakeholder. (See Phase Two: Determine Resources, Key roles on a data project team.)
- Driving deep stakeholder engagement in decisions regarding cross-functional challenges and priorities, even when challenging. (See Phase Four: Tune the Change, Learner.)
- Making trade-offs, listening to diverse perspectives, knowing when to be willing to be misunderstood and investing in relationship repair. (See Chapter 6, Tune the Change, Level three: learner.)
- Being ready to adjust along the way as conditions continue to change (See Phase Four: Tune the Change) helps others adapt by assisting them to challenge current mental models to adopt new ones. (See Chapter 1, Data Foundations, The human side.)

Data helps early detection of risks

Identifying risks as part of the project-change approach gives the project lead an opportunity to have extra meetings with a sponsor experiencing doubts, a group member that has had a change in their commitments or anyone involved with a purchase order falling through the cracks in procurement. Detecting risks early can help prevent setbacks in confirming commitment, securing resources and getting funding approval.

Setting up repeatable processes and supporting structures will enable you to scale your efforts and, from there, fine-tune your efforts (see Table 6.16). Thinking about what can go wrong before the project starts (see Table 3.9) is time well spent, increasing awareness of what could go wrong with your investment.

The summary in Table C.2 applies computational thinking from Chapter 1 (see Table 1.6). Identifying risk is something a lot of project teams skip. The list in Table C.2 is not exhaustive, but displays some common areas of risk we've reviewed.

Table C.2 Summary of common risks and mitigation tools

Questions to consider	Mitigations/Resources
Is the end deliverable clear? Measurable? Were the right stakeholders included for added perspective?	Run through pre-implementation review and thoroughly review the five factors for scoping: clear deliverable, critical success factors, critical success criteria, key performance indicators and ongoing feedback. *Appendix E: Pre-Implementation Review and Post-Implementation Review Templates* *Worksheet 5: Scoping Template*
Do the groups impacted have the desire for the shift necessary? Are the from–to vision and objectives clear?	Confirm understanding of sponsor and group member expectations for change before, during and after the shift. *Worksheet 9: PCT Assessment Summary and Risk Profile*
Do the groups impacted have the necessary skills to execute the project?	Confirm the business objective against available skills when selecting sponsors, group members, SMEs and partners. *Worksheet 10: PCT Sponsor Assessment and Worksheet 11: PCT Group Member Assessment*
Think through competency needs	*Worksheet 16: Augment Core Competencies (Learn)*
Do you own all the data needed to realise internal efficiency goals? External business goals?	Confirm (or secure) the necessary legal rights to the data for what you want to do today and potentially in the future. *Worksheet 17: Data Supply Chain: Core Systems*
Will the architecture support your project's objectives? Can you go open source or not? Will your environment be virtualised in the cloud or not? Will your governance efforts be centralised, federated or decentralised? How should you manage external data warehousing to avoid unexpected costs?	Understand the many risks associated with making architectural choices. Vet implementation challenges. *Worksheet 4: Project Vetting Checklist* *Worksheet 15: To-Dos List (Practice)*

Vetting your data needs – with rigour

Creating a thorough inventory of data needs is critical to your data project's success. It provides a practical understanding of business priorities, infrastructure needs, legal and ethical concerns, data governance needs and, ultimately, business value and impact.

Your project inventory should include the following (see applied practices from Chapter 1, Data Foundations, The machine side):

- Classification of the data needed (name, definition, type, source, ownership and so on).
- Grouping data types into categories helps reduce the overall data types you will work with.
- Usage need. (What level of granularity, collection frequency and so on; how long do you have to keep it for?)

Becoming sensitive to the change impact

Understanding the project scope helps identify what is needed to achieve objectives. Only then can you start to plan for the mindset shifts and related behaviours that need to happen. Thoroughly understanding both project and change goals helps to calibrate the speed of change you can operate at to maintain success and keep stakeholders on board. Too project-focused, too transactional, and you risk losing support. Too slow, too relational and accommodating, and you risk scope creep or worse, or eroding goals.

One aspect of any data project that is important to recognise has to do with people's fear of how automation might impact their role. It is important to reflect on this and other concerns that come up by communicating, early and often, the intent for the change, how it will impact people's tasks and roles, the benefits it will bring and what opportunities are created as a result (see Chapter 1, Data Foundations, The human side).

> *Tip*: Adopting a data-driven approach by using analytics to measure and understand the change efficiency and impact helps to model the change behaviours you are looking for others to adopt. Sentiment analytics helps reflect how change is being perceived. You'll want to reflect on how change is *actually* happening and whether there are specific change roles that are more efficient than others. What behaviours are they enacting that others are not?

PITFALLS TO AVOID

Many data efforts suffer because they were approached incorrectly. Here are several familiar challenges. Learn what to avoid and how you might navigate them better.

Lack of data intelligence in leaders or stakeholders

With data now elevated through the creation of the CDO role in organisations, and by quickly becoming an essential fourth pillar of improving organisations (people, process, technology platform), data intelligence is fast becoming a necessary workforce skill. Data intelligence impacts all levels across all sectors. Everyone needs to develop data skills to determine data quality and confirm data sources, effectively influence decision-making and use data responsibly and ethically.

Cultivating data intelligence starts with observing the current mindset and establishing a from–to state. We consume data in our lives and work like we consume news and other information. Because of that, we need to learn to do it critically, responsibly and ethically.

A common assumption occurs when a CDO is hired and their data organisation is staffed: their presence will check the box for the organisation to utilise data intelligently as an ongoing resource. The CDO will sponsor data projects that will provision data through a common architecture, provide insights from data scientists, collaboratively govern the use of data and upskill stakeholders with tools and training – to name just a few things. However, data intelligence is *everyone's* responsibility. Investing in improving the data architecture, analytics, governance and enablement of data is the first step towards extracting the full value of the data supply chain. Effectively partnering with human resources to infuse their mechanisms with data-forward language helps to guide the broader workforce towards ongoing digital upskilling.

If executives do not model transparent decision-making using data, if employees do not integrate data into their decision-making and if those in charge of performance management do not set expectations and reinforce the need to apply data across all functions, all the reporting and dashboard efforts won't be fully utilised. Insights won't be put into action.

Many organisations believe that staffing a team to generate reports means data is driving the business. The fact is, there is very little training and enablement to teach data users how to interpret those reports, how to ask follow-up questions and how to put the data into action. There is even less training for managers of teams to aid in reinforcing this discussion.

Tip: Infuse manager support mechanisms with data literacy tools to help managers become better advocates for driving data awareness and ongoing usage of data tools for ongoing decision-making. Managers are the organisation's primary carriers of culture and determine its effectiveness. They create culture by setting clear targets, expectations and objectives while encouraging employees to grow and develop in a trusting environment.[95] They cascade executive communications, administer human resource policies, reinforce organisational tenets and keep the pulse of work execution. Lack of manager-level understanding and buy-in for integrating data skills represents a tangible barrier to capturing the full potential of the company and could lead to further estrangement of the data team's efforts.

---

95 https://www.gallup.com/workplace/236555/managers-create-high-performance-cultures.aspx

Perception that machine learning and artificial intelligence is a black box

There is nothing magical about data, mathematics and algorithms. Formulas perform how they should. Machines do what they are told. The less transparent aspect of data is causes, correlations and meaning. Machines don't learn by themselves; they must be *taught* to interpret data.

Overestimating what machine learning, artificial intelligence and even automation (to some degree) can do for the organisation can create a huge expectation gap – for everyone. Such misunderstandings could result in legal, trust and reliability issues impacting financial performance.

> *Tip*: Always remember to educate and caution stakeholders that artificial intelligence is not an oracle but a way of applying advanced algorithms to vast amounts of data. Artificial intelligence is a *convenience*.

Perception that task automation means the death of jobs

There is a very real fear among many that data and technology will automate enough tasks to eliminate human workers from the workplace. There are certainly use cases in practice. Ultimately, automation is generally used to replace highly repetitive jobs which some argue require little skill, such as warehouse work, retailing, proofreading, editing, bookkeeping, courier services, market research analysis, long-haul driving, production, home care work, sales and manufacturing. There is little that will not be impacted by automation.

Yet, a Deloitte study of automation in the UK found that while 800,000 low-skilled jobs were eliminated due to AI and other automation technologies, 3.5 million new jobs were created, paying on average nearly $13,000 more per year than the ones lost.[96] Workers who can work *with* machines are more productive than those without them; this reduces both the costs and prices of goods and services. Over the long term, creativity (that is, developing and knowing how to wield dynamic learning skills and principles, critical thinking, exponential thinking, creative problem-solving and so on) is what will distinguish humans from machines.

Managing and monitoring data models, automating reporting efforts, evaluating ethics of algorithm decision-making, driverless cars and AI-powered search are all examples of how we will start to work better with data and machines. We are, in essence, learning to manage machines that perform our original tasks.

---

[96] https://www2.deloitte.com/content/dam/Deloitte/uk/Documents/Growth/deloitte-uk-insights-from-brawns-to-brain.pdf

Tip: Automation is only part of the value of data and machines. Using it as the primary benefit undermines all the benefits from the prior section (and this book!). Balance the benefits that best align with your organisational culture and share which enables *employees and the organisation* to evolve beyond current performance.

Perception that everyone is an analyst

When evangelising data fluency skills, don't undervalue the skills of data scientists. Declaring everyone an analyst means no one is an analyst. Isolated pockets of analytics in random parts of the organisation do not enable the data scientists or anyone else to connect with the data. When data scientists are working on high-impact problems, but no one knows about their work, accountability to act on their findings remains vague. Lack of accountability in using data insights to drive decisions means anecdotes are enough to refute actual data.

These examples are symptoms that leadership doesn't fully understand, buy into or support the value of data science. They show that data science is there to ensure the company 'has data', not necessarily that it 'uses data'.

Tip: Improving the maturity of data intelligence of an organisation follows the data maturity of its executive leaders and managers. Leaders at every level must understand, value and constructively use data for the rest of the organisation to care about adopting data-forward behaviours. This directly impacts data projects and data science efforts' ability to scale.

Perception that dashboards are the end goal for data initiatives

Understanding and valuing data is one thing. Making the most constructive use of data and data resources (like data scientists) is another. Most employees' (and leaders') experience with and exposure to data is through reports and dashboards. This leaves many of us to assume dashboards are the main output of data operations projects – that the main effort is merely to automate those burdensome manual reports tying everyone down. This assumption is so common and a primary reason many data science efforts fail.

Remember the need for exponential thinking from Chapter 1, where innovative technology is first used to generate more of what it already has? Automation is essential, but the main goal of ML and AI is to help the organisation do more of what it is already doing. Data science, being a forward view on data, is to help the organisation move *beyond* what it is already doing.

With automation, the starting point is the data consumer and the dashboard. To be truly data-driven, the starting point must be the data itself. The data indicates what you need to look at and act on; analysis is predictive so the organisation can become proactive.

For data science to create real value for companies, there must be a complete end-to-end understanding and view of how their projects will be used in production. However, often, appropriate consideration is not given to this aspect. This can happen because data science teams do not have an architectural view of how their projects will be integrated within the production pipeline since different teams typically manage these projects. Likewise, those teams share a limited view of the data science development process and how a soon-to-be-developed project will fit into their environments. This misalignment can often result in ad hoc data science projects that don't deliver business value.

> *Tip*: Organisations must leverage data science automation to gain greater agility and faster, more accurate decision-making. To do this, functional leaders must always aim to become the silo-busters their teams need, so the development-to-production view is transparent to all involved. The emergence of AutoML[97] platforms allows organisations to become more agile by enabling them to utilise current teams and resources without recruiting additional talent. AutoML 2.0 empowers BI developers, data engineers and business analytics professionals to leverage AI/ML and add predictive analytics to their BI stack quickly and easily. By providing automated data pre-processing, model generation and deployment with a transparent workflow, AutoML 2.0 is bringing AI to the masses, accelerating data science adoption.

Using dashboards by themselves assumes all the questions are known from the start. Report automation and data science combined elevate the dashboards' role. Using them in conjunction helps data consumers learn to ask better questions as they become increasingly fluent in what the historical data tells them. Thus, combined insights enable teams to monitor and learn from ongoing preventative actions and to make long-term strategic decisions.

Perception that data (including ML or AI) is inherently ethical and objective

That human conceptions of race, gender, social class and other discriminatory characteristics are consciously or unconsciously built into data models and algorithms is well documented.[98] [99] We would like to think that we are unbiased if we are socially aware and educated. However, we all have biases towards one thing or another.[100] Inadvertently integrating those biases into self-learning algorithms can have profound consequences on the performance of the organisation's algorithms.

The machine's learned bias is only one part of it. The use of personally identifiable information (highly valued by advertisers and law enforcement) and the transparency/explainability of how AI insights are derived is now part of the EU's GDPR (General Data Protection Regulation).[101]

---

97 Automated Machine Learning (AutoML) automates the training, tuning and deployment of machine learning models. AutoML can be used to automatically discover the best model for a given dataset and task without any human intervention.

98 Race and gender: http://proceedings.mlr.press/v81/buolamwini18a/buolamwini18a.pdf

99 Social inequity: https://compass.onlinelibrary.wiley.com/doi/full/10.1111/soc4.12962

100 https://www.americanbar.org/groups/business_law/publications/blt/2020/04/everyone-is-biased/

101 Explainable AI, or Interpretable AI, or Explainable Machine Learning, is artificial intelligence in which humans can understand the decisions or predictions made by the AI. It contrasts with the 'black box' concept in machine learning where even its designers cannot explain why an AI arrived at a specific decision. https://en.wikipedia.org/wiki/Explainable_artificial_intelligence

Tip: Ethics must be considered at every aspect of the data supply chain – from discovery through consumption. It's also something to consider when generating business models and working methods and in how teams are formed – the functional frame we bring to conversations has inherent blind spots. This is why collaboration and consensus are so critical (see Phase Two: Determine Resources, Designing the Project Team). Of course, avoiding lawsuits is important, but securing sustainable, trustworthy data for your business is more important. Ultimately, it makes your data asset more valuable.

Perception that data is free (of cost and legal obligation)

Data does not want to be free; it wants to be curated, governed and used. Recalling the water/utility and raw crude/refined oil examples, raw materials must undergo multiple transformation processes to provide the organisation with current, future or potential (ongoing) business value.

Likewise, data is not free. It requires a significant, ongoing investment in specialised staff, efficient processes and technology platforms that deliver an ethical data supply chain. It also requires a workforce digital upskilling investment, enabling internal and external data users to become more fluent and effective in their roles.

Data fluency includes increasing awareness of data privacy, security and legal use of data. Even if there are no plans to monetise the data, employees need to be generally aware of how the General Data Protection and Regulation (GDPR) impacts their business model design and value.

Tip: Regardless of how data is used, it is important to have the legal rights to use the data. Address this issue early in developing your data strategy to avoid unintentional violations. Lack of attention to legal rights to use data can result in confusion of ownership or being prohibited from selling your new product or solution because it is using data you were not authorised to use.

Perception that changes will be quick because they are logical

Taking time to think through the change you are looking to create with your data project is essential to success. Understand the population that will need to shift and develop personas. Think through multiple change scenarios for the business when introducing a data strategy or new data initiative.

Common mistakes related to managing change:

- There is not a strong sponsor model in place. (Phase Two: Determine Resources)
- The sponsor doesn't know how to be a sponsor. (Phase Two: Determine Resources)
- Underestimating the scope of the change. (Phase One: Scope the Project)

- Not taking time to plan through what must happen. (Phase Three: Manage the Work)
- Failing to understand that graduating to predictive analytics means reporting must be more intentional; questions are not known from the start as they are with dashboards. (Phase Four: Tune the Change)
- Common story telling mistakes that over emphasise progress metrics as opposed to business impact when considering the increased value of the data asset. (Phase Four: Tune the Change)
- Focusing purely on cost efficiency of operational changes versus overall business value undermines data strategy and value of the data asset. (Data Foundations, Phase One: Scope the Project)
- Not understanding or measuring the impact of operational improvements. (Chapter 1, Data Foundations)
- Making such minor changes that nothing changes. (Phase One: Scope the Project)
- Forgoing or underestimating the need for communication of the change and just moving forward. (Phase One: Scope the Project; Phase Two: Determine Resources)

Forgetting baseline measures to benchmark value

Introducing new data initiatives focuses on future measurements. This can prohibit early success tracking if we forget to benchmark the current state.

Additionally, scrapping the old to bring in the new means the company cannot statistically prove the value of having taken a new step. Technology platforms' rip-and-replace and upgrade-and-embrace strategies have influenced strategic thinking in other functions to think about the one-stop-service model.

Tip: Integrating agile development and ongoing release delivery enables the ability to increase adaptability and more fluid movement between what has worked in the past and what is needed to succeed in the future.

ANNEX
PROFESSIONAL RESOURCES

Business intelligence and data management blend business and technology. They facilitate collaboration and communication between disciplines and across sectors.

Professional organisations unite business intelligence students and practitioners through networking opportunities, online communities and annual events. They also offer job listings and career advice, continuing education programmes and news updates about the business intelligence field.

Artificial Intelligence Risk Management Framework (AI RMF 1.0)

As directed by the National Artificial Intelligence Initiative Act of 2020 (PL 116-283), the goal of the AI RMF is to offer a resource to the organisations designing, developing, deploying or using AI systems to help manage the many risks of AI and promote trustworthy and responsible development and use of AI systems. The Framework is intended to be voluntary, rights-preserving, non-sector-specific and use-case agnostic, providing flexibility to organisations of all sizes and in all sectors and throughout society to implement the approaches in the Framework. This publication is available free of charge at: https://doi.org/10.6028/NIST.AI.100-1

Association to Advance Collegiate Schools of Business

AACSB International unites business students, educators and professionals from around the world to improve the quality of business education. Members receive association publications, access to career resources, and accreditation updates. AACSB International also holds more than 90 conferences and seminars annually.

BCS, The Chartered Institute for IT

BCS, The Chartered Institute for IT, is a professional body and a learned society that represents those working in information technology (IT) and computer science, both in the United Kingdom and internationally. The organisation offers resources and certifications in digital skills to raise trust in the IT profession and ensure IT is used effectively and ethically to solve the biggest problems of society.

Control Objectives for Information and Related Technologies (COBIT)

COBIT is an IT governance framework for businesses wanting to implement, monitor and improve IT management best practices. Today, COBIT is used globally by all IT business process managers to equip them with a model to deliver value to the organisation and to enable them to apply better risk management practices associated with the IT processes.

Data Management Association (DAMA)

DAMA is a non-profit and vendor-independent association of business and technical professionals that is dedicated to the advancement of data resource management (DRM) and information resource management (IRM). The DAMA-DMBOK describes key principles such as data governance, data architecture, metadata management, reference and master data management, data security, data quality, data ethics and other important foundational elements that data programmes need to incorporate to be successful.

Digital Analytics Association

Dedicated to building a better digital world through data, the DAA seeks to standardise practices, terminology and certification in the field. The DAA offers professional development, career resources and networking opportunities to members.

EDM Council

EDM Council is the member-driven, neutral, global trade association dedicated to elevating data management and analytics as a strategic business priority and providing best practices, standards and education to data and business professionals. The EDM Council focuses on a data capability assessment model (DCAM) and defines the scope of capabilities required to establish, enable and sustain a mature data management discipline. It addresses the strategies, organisational structures, technology and operational best practices needed to drive data management successfully.

International Institute for Analytics

IIA helps business leaders foster effective analytics. Member benefits include advisory services, training and research networks. The IIA empowers analytics leaders to grow by providing unbiased data and assessments to healthcare, manufacturing, finance and retail professionals.

International Institute of Business Analysis

The IIBA supports the expanding business analysis field through certification and professional development programmes. Members gain access to events, mentorship opportunities and business data analytics resources. Other available benefits include quarterly publications, webinars and best practices content.

National Black MBA Association, Inc.

The NBMBAA promotes diversity and inclusion in the business world by advocating for underrepresented students, entrepreneurs and professionals. With 39 local chapters in North America, NBMBAA offers online resources, including webinars, professional certification courses and pillar programmes. NBMBAA's three pillars emphasise education and career, leadership and professional development and entrepreneurship.

Project Management Institute (PMI)

PMI services include the development of standards, research, education, publication, networking opportunities in local chapters, hosting conferences and training seminars and providing accreditation in project management. The Project Management Body of Knowledge (PMBOK) is a set of standard terminology and guidelines for project management. The body of knowledge evolves over time and is presented in *A Guide to the Project Management Body of Knowledge*, a book whose seventh edition was released in 2021.

ProSci® Change Management (ProSci®)

The ProSci Change Management methodology is a systematic and holistic approach that guides organisations to realise the benefits of their change initiatives. It also aims at building internal, organisational capabilities to deal swiftly and efficiently with the ever-increasing number of changes. The Change Management Prosci® methodology is built around two fundamental models: one developed to deal with change at a personal level and the other to deal with change at an organisational level. See https://www.prosci.com/

The model for personal change is based on ADKAR®, an acronym that stands for Awareness, Desire, Knowledge, Ability and Reinforcement.

The model for organisational change is realised through the Three-Phase Model with defined steps and tools for every phase: Prepare Approach, Manage Change and Sustain Outcomes.

APPENDICES

APPENDIX A
DETERMINING MULTIYEAR GOALS

In Phase One: Scope the Project we discuss how to identify your project needs according to your multiyear goals. If you don't have a good grasp of your multiyear goals, you can use the sample goals in Table A.1 as a starting point. How relevant or important are these common goals to you? How would you measure your progress against the ones you do think are important? If you can think of other goals, include them, and some possible metrics, in this list. When you're done, you can ask your executive team for their thoughts on the list.

Table A.1 Determining multiyear goals

GOAL	SAMPLE MEASURES OF SUCCESS
Increase the number of stakeholders we serve	1. Total stakeholders served
	2. Percentage of organisation in need of data services
Increase the impact on each stakeholder we serve	3. Percentage of stakeholders who report achieving the intended outcome from data effort (such as monitoring diagnostics, reporting insights or influencing decisions)
	4. Result of satisfaction surveys completed by stakeholders
Have an impact on policy, process or result addressing key measures of team (increase revenue, manage costs, increase quality, etc.)	5. Percentage of relevant team or organisational policies aligned with data service needs that are implemented (and followed regularly)
	6. Number of senior stakeholders advocating data service needs in their functional strategies
	7. Number of teams or functional business units consuming data from common data sources
	8. Increase of funding for data projects
Develop more sustainable sources of funding	9. Number of functional leaders funding data service efforts
	10. CFO partnership +/or primary sponsorship
Reduce cost per outcome	11. Number of people served while budget stays flat
	12. Cost of communications, training and ongoing enablement

(Continued)

Table A.1 (Continued)

GOAL	SAMPLE MEASURES OF SUCCESS
Develop and increase the role of data champions in the organisation	13. Number of champions per business unit
	14. Data service projects affecting all internal functions – programme, administration, budgeting, etc.
	15. Percentage of all labour needed to run the data services effort
Become an employer of choice, the team to work for	16. Number of résumés received for each job posted
	17. Percentage annual turnover
	18. Number of requests for informational interviews from leadership to line level

APPENDIX B
POTENTIAL DATA PROJECTS

Developing a data strategy will require the project manager and stakeholder groups, as governing committees, to rank or prioritise the business measures of the organisation to support ongoing management, storage, analytics efforts (and justify ongoing maintenance costs).

Data category	High-level tasks	Deliverables
Governance process and framework	Document the governance framework and the policies to help enforce the future state data quality framework. Development of a high-level conceptual governance framework for ongoing management of change requests. Document the change management processes to manage future modifications to metrics, hierarchies and calculation requirements.	Draft future state governance structure Change process documentation
Assess metrics/data	Using reporting team to scope, documentation of current state of metrics, calculations. Document the data definitions of the metrics in the scorecards in scope. Assess the gaps and inconsistencies in definitions. Confirm change request process.	Metrics definitions documentation (dictionary) with complete metadata Updated change process Master data standards guide
Assess data hierarchies	Confirm geographic, product and financial hierarchies. Confirm hierarchy related to scorecards/reports. Recommend change request process for hierarchy changes. Recommend policy for hierarchy changes (what changes should we (not) accept and why (not)).	Hierarchy definitions documentation Hierarchy change policy

(Continued)

Data category	High-level tasks	Deliverables
Assess data tools	Confirm tools roadmap. Determine relationship of manual to automated (related to metric scope). Assess quality issues.	Input into tools strategy – process? Recommendation for tools consolidation? If outsourcing, opportunity for contract review and/ or consolidation?

APPENDIX C
WORKING AGREEMENTS, DATA CONTRACTS AND SERVICE-LEVEL AGREEMENTS

TEAM WORKING AGREEMENTS (PRODUCED BY PROJECT MANAGERS)

A working agreement is a document that lists out a team's norms using statements like 'We value face-to-face interaction over long-form asynchronous communication'. These documents are a way to take implicit agreements and make them explicit so that the team can remain aligned throughout the duration of a project. This can be a particularly useful tool when a new team is spinning up and norming together.

Working agreements are so valuable that all project management methodologies (that is, 6Sigma, PMP, Agile and so on) have a template for forming working agreements. The trick is remaining accountable to them when it is difficult to do so.

Common elements in working agreements include:

- *Working hours*: When you will meet and for how long. All team members agree to be available during those hours.
- *Phones*: If team members can have them or not, and any exceptions for this rule.
- *Product owner*: Their availability and the ways of contacting them in case of an emergency.
- *Definition of complete*: When the team can agree on the task being finished, and they are ready to move on.
- *Respect*: Making sure each team member always has a chance to share their opinion.
- *Absence*: What happens when the team member is not present during a sprint. A common practice is to agree that we support decisions agreed on during our absence.

DATA CONTRACTS (PRODUCED BY DATA ENGINEERS)

In an ideal world, data engineers are given a direct connection to data from source systems. Instead of simply extracting the data from the source system on their own, a data engineer should work with the source system team to establish a data contract. The contract can be implemented in many forms and should be stored in a system that is both easily discoverable in the organisation and can be integrated into the development process programmatically.

The properties of your data contract may vary. For example, instead of an ingestion frequency you may want to store some other form of schedule notation. The point here is to document the fact that the ingestion exists, what system it impacts, how it impacts that system (frequency, volume and so on), and clear ownership of both the source system and the ingestion job.

SIMPLE SERVICE-LEVEL AGREEMENTS FOR INTERNAL DATA PROVISIONING

A data-related SLA is an agreement between a data team and their stakeholders that covers all the key infrastructure elements and service metrics such as data ownership, accuracy, completeness, service uptime and network availability.

For example, a company can draw up an internal SLA between its engineering department and its marketing team. This SLA might specify that marketing needs to provide a certain number of leads to marketing per month to help them reach their quota for an event.

The SLA helps mitigate the confusion, miscommunications and unfilled commitments that can occur between customers, service providers and internal departments for multiple reasons. Such setbacks can help set clear expectations from the start of any business relationship.

APPENDIX D
TYPES OF DATA PROJECTS

Examples of data operations projects

Project	Metadata management	Data warehousing	Data quality and lineage	Balanced scorecard management	Data strategy
What is it?	Metadata is data about data (such as time stamps, location, job execution logs and data owners). Data documentation should be created and maintained to ensure it can be understood in the future by those not involved in the project when first implemented. Consider what information you would need to understand a dataset if you accessed it in ten years' time.	Data warehousing integrates data and information collected from various sources into one comprehensive database. For example, a data warehouse might combine customer information from an organisation's point-of-sale systems, mailing lists, website and comment cards.	Data quality is a measure of the condition of data based on factors such as accuracy, completeness, consistency and latency. Data lineage is a technique used to help troubleshoot quality issues where data is coming from, where it is going and what transformations are applied to it as it flows through multiple processes. It helps understand the data lifecycle.	A balanced scorecard (BSC) is defined as a management system that provides feedback on both internal business processes and external outcomes to continuously improve strategic performance and results. Cascading reports support strategies at each level of the business.	A data strategy is a long-term plan that defines the technology, processes, people and rules required to manage an organisation's data and information assets. Generally, it includes plans for identifying top business questions and prioritising KPIs, developing an ongoing business requirements process for reporting and a single architecture to support the reporting scope. Easily connects to ongoing governance efforts.

(Continued)

Examples of data operations projects (Continued)

Project	Metadata management	Data warehousing	Data quality and lineage	Balanced scorecard management	Data strategy
Who is responsible?	A data project lead assembles and defines roles for business data stewards (BDSs). BDSs manage the business metadata, including glossary terms and bus ness rules, providing support for generating and consuming metadata. System data stewards manage the technical metadata and the alignment of systems with business rules.	A data project lead works with data warehousing specialists (DWSs) to create a solution for improving efficiency and reducing costs. DWS professionals develop processes and procedures for managing data across an organisation or within a specific project. They design software applications or programs to handle tasks related to data storage and management.	A data project lead collaborates with data quality managers, analysts and engineers who are primarily responsible for fixing data errors and other data quality problems in organisations, to summarise data quality issues and determine next-step actions.	A data project lead collaborates with functional leaders and analysts to define, prioritise and map key performance indicators across four perspectives (financial, customer, internal process and learning and growth) to the overall business strategy.	A data project lead aligns data team strategy with needs of business managers. Together they review data needs, help prioritise and manage data operations to support ongoing reporting needs and ensure data strategy aligns with overall business strategy and regulatory requirements.

(Continued)

Examples of data operations projects (Continued)

Project	Metadata management	Data warehousing	Data quality and lineage	Balanced scorecard management	Data strategy
Why is it important?	Metadata management increases trust and accuracy of data by becoming a single source of truth. For a large volume of complex data coming from many sources MDM enables the ability to find, understand and trust the right information to gain actionable insights that improve business operations.	Data warehousing improves the speed and efficiency of accessing different datasets and makes it easier for corporate decision-makers to derive insights that will guide the business and marketing strategies that set them apart from their competitors.	Data is used to aid in decision-making processes, which has led to an increased importance of data quality in a business. Data quality ensures that the information used to make key business decisions is reliable, accurate and complete.	BSC enables companies to better align their organisational structure with the strategic objectives and KPIs. To execute a plan well, organisations need to ensure that all business units and support functions are working towards the same goals (and adopt a collective understanding of KPIs).	A data strategy helps identify the best talent, processes, technical tools and datasets that meet business needs and support both data teams, business stakeholders and extended users. The strategy validates/ verifies that the talent, processes, technical tools and datasets meet all data governance policies, ensuring compliance with regulations and moral alignment with the company's mission.

APPENDIX E
PRE- AND POST-IMPLEMENTATION REVIEW TEMPLATES

PRE-IMPLEMENTATION REVIEW TEMPLATE

	Project failures
	Use this space to detail ALL the potential ways this project failed – especially the kinds of things the team ordinarily wouldn't mention as potential problems, for fear of being impolitic.
1	
2	
3	
4	
5	

Thinking through potential failures and mitigations is the start of a Risk Register that can be managed and tracked throughout the project. The most important aspect of this exercise is that it's done as a team. Research other risk registers online, and adapt as needed.

Failure	Risk theme	Rank (severity)	Potential mitigation	Still in scope?
1				
2				
3				
4				
5				

POST-IMPLEMENTATION REVIEW TEMPLATE

Project vitals			
Project title			
Project lead			
Supporting team members			
Executive sponsor		Estimated value/Impact	
Project focus			
Start date		End date	
Project challenges	Did the project finish on time? Were there any changes in headcount/resourcing during project work? Were there any changes in scope during project work? Use this space to describe any major difficulties encountered.		

Project description	
Project title	
Situation	
The approach	
The impact	
Major phases of work	
List of key deliverables	
Other notes	

APPENDIX F
DATA TERMS, JARGON AND PHRASES

Data has become the forefront of many mainstream conversations about technology. New innovations constantly draw commentary on data, how we use and analyse it, and the broader implications of those effects. As a result, popular IT vernacular has come to include several phrases, new and old:

- *Analyst*: A professional who is responsible for examining, studying and interpreting data, information, or various forms of evidence to derive insights, make informed decisions or provide recommendations.

- *Analytics*: Analytics uses data to answer business questions, discover relationships, predict unknown outcomes and automate decisions. This includes finding meaningful patterns in data and uncovering new knowledge based on statistics, predictive modelling and machine learning techniques.

- *Anonymisation*: A data-processing technique that removes or modifies personally identifiable information; it results in anonymised data that cannot be associated with one individual. It's also a critical component of most organisations' commitment to privacy. Using unique identifiers to avoid duplicate data can sometimes increase the risk of de-anonymising the data. While it is not always possible or guaranteed, it's vital to be sure you respect the informed consent given to you by the original data disclosers.

- *Artificial Intelligence*: Artificial Intelligence (AI) refers to the development of computer systems that can perform tasks that typically require human intelligence. These tasks include learning, reasoning, problem-solving, understanding natural language, speech recognition and visual perception. AI aims to create machines that can mimic cognitive functions and, in some cases, surpass human capabilities in specific domains.

- *Big data*: Big data refers to large and complex datasets that are beyond the capacity of traditional database systems. Big Data is characterised by its volume, velocity, variety and complexity, requiring specialised tools and technologies for storage, processing and analysis.

- *Big data analytics*: The process of collecting, organising and synthesising large sets of data to discover patterns or other useful information.

- *Business intelligence*: The processes, technologies and tools used to collect, analyse and present data to support business decision-making. BI systems help organisations transform raw data into meaningful insights and reports, enabling informed strategic choices.

- *Database*: A collection of data points organised in a way that is easily manoeuvred by a computer system.

- *Data centre*: Physical or virtual infrastructure used by enterprises to house computer, storage and networking systems and components for the company's IT needs.

- *Data integration*: A set of capabilities that allow for synchronisation of data across multiple data producers and make rationalised data available to consumers. Data integration tools are selected based on temporal needs, data volumes and transformation complexity.

- *Data integrity*: The validity of data, which can be compromised in a number of ways including human error or transfer errors.

- *Data lakes*: Centralised repositories that store vast amounts of raw, unstructured and structured data at scale. Data lakes provide a flexible and scalable storage solution, allowing organisations to store and analyse diverse datasets.

- *Data miner*: A data miner is a professional who uses various techniques to analyse large sets of data, extracting valuable patterns, information and insights. They employ data mining methods to uncover trends, relationships or hidden knowledge within datasets, helping businesses and organisations make informed decisions based on the discovered patterns.

- *Data mining*: Data mining is the process of discovering meaningful patterns, trends and insights within large datasets using various techniques and algorithms. It involves extracting valuable information from data to aid in decision-making and uncover hidden relationships or patterns.

- *Data privacy*: A sub-discipline within data security concerned with the proper handling of data – consent, notice and regulatory obligations. Typically, data privacy concerns consist of:

  o Whether or how data is shared with third parties.

  o How data is legally collected and/or stored.

  o Regulatory mandates such as GDPR, HIPAA, GLBA or CCPA.

- *Data quality*: The critical practice of checking, correcting, labelling and normalising data so that accurate analysis can be done later. It includes activities such as checking for accuracy (consistent formats for date and time) or determining or creating a unique identifier (such as an email address or phone number, allowing the combination of one dataset with others), to ensure accuracy and quality in each stage of the data supply chain.

- *Data retention and business continuity*: Pertains to archiving of data in Tier II and/or Tier III storage to keep the footprint on the core transactional system nimble and performant. The reasons for storage are driven by business needs and regulatory compliance mandates. Also, periodic backups are done to maintain business continuity by safeguarding against unexpected data loss.

- *Data rivers*: Real-time or near-real-time data streams that flow continuously and are processed as they arrive, allowing for quick analysis and decision-making.

- *Data security*: Consists of policies, procedures, controls and technologies to secure data at rest and in transit across all applications and environments (production and non-production).

- *Data taxonomies*: Data taxonomies are hierarchical structures that classify and organise data elements based on their characteristics, facilitating efficient data management and retrieval. They provide a systematic way to categorise and organise data, helping users understand relationships between different data elements and improving overall data governance.

- *Data warehouse*: A data management system that uses data from multiple sources to promote business intelligence.

- *Little data*: Little data refers to a small-scale dataset that is typically manageable and can be easily handled using traditional data processing tools. It is characterised by a volume of data that doesn't require advanced processing or storage capabilities.

- *Machine Learning*: A subset of artificial intelligence that focuses on developing algorithms and models that allow computers to learn from and make predictions or decisions based on data. It enables machines to improve their performance on a specific task over time without being explicitly programmed.

- *Metadata*: Summary information about a dataset, such as its name, definition, calculation, source name, business owner and so on.

- *Principle of least privilege (PoLP)*: An information security concept which maintains that a user or entity should only have access to the specific data, resources and applications needed to complete a required task.

- *Raw data*: Information that has been collected but not formatted or analysed.

- *Structured data*: Any data that resides in a fixed field within a record or file, including data contained in relational databases and spreadsheets.

- *Unstructured data*: Information that does not reside in a traditional column–row database like structured data.

WORKSHEETS

Worksheet 1 - Building a vision
Worksheet 2 - Reflection: recall your past data project experiences
Worksheet 3 - To-do list based needs assessment
Worksheet 4 - Project vetting checklist
Worksheet 5 - Scoping template
Worksheet 6 - Change management preparation task list
Worksheet 7 - Assessing your organisation for change
Worksheet 8 - Integration of project and change management activities
Worksheet 9 - PCT assessment summary and risk profile
Worksheet 10 - PCT sponsor assessment
Worksheet 11 - PCT group member assessment
Worksheet 12 - Mapping discovery at the end of Phase Three
Worksheet 13 - Mission and guiding principles (observe)
Worksheet 14 - Three hows – multiyear goals (practice)
Worksheet 15 - To-dos list (practice)
Worksheet 16 - Augment core competencies (learn)
Worksheet 17 - Data supply chain: core systems

WORKSHEET 1
BUILDING A VISION

Use this example and the following template to assemble a draft vision for your data project. It is helpful to revisit progress against this vision every quarter.

Dynamic learning in action	
Use the following to practice short- and long-term thinking. Look at the example of a team's goals. Notice how the Landscape issues influence the vision.	
Vision	**Landscape**
Team's North Star — Drive best-in-class, end-to-end experiences and processes across the organisation. Working backward from our customers-stakeholders, build mechanisms that enable us to increase (success metric: revenue, quality, cost savings, etc.) at scale. • *Develop (processes, programmes, products, services) goal here.* • *Deliver (processes, programmes, products, services) here.* • *Use data and research insights to continuously improve customer-stakeholder engagement to increase (processes, programmes, products, services) (success metric: revenue, quality, cost savings, etc.).* *Year 3: The team will automate 100% of (processes, programmes, products, services), optimising year over year efficiency and cost savings goals.* • *100% of annual financial audit will be automated.*	• *100 roles covered by FY20xx; represents 85% of total headcount.* • *Need faster insights and health diagnostics.* • *Engage stakeholders to utilise and fully implement (processes, programmes, products and services).*

- *100% of service tickets will be routed to IT as part of a joint service-level agreement.*

- *All stakeholders will have real-time insights and monitoring of health diagnostics for a complete end-to-end process.*

Work backwards plan

Year 2:

- *Using third-party platform to determine 'on the fly' solution construction.*

- *Deploy 'Voice of the Customer' surveys to ensure we are meeting stakeholder and end-customer needs.*

Year 1:

- *Develop and deploy real-time solutions analytics to our project team.*

- *Continue process adoption and stakeholder buy-in.*

- *Invest in solid documentation (how-tos, user guides, PM playbooks, etc.).*

- *Continue process automation expansion via internal platform to expand ticketing coverage to 100% of all financial audit tasks and ticketing tasks.*

- *Deploy customer feedback addendum to service tickets.*

- *Develop stakeholder feedback report.*

- *Deploy language choice/filter once translations are available.*

Success measures	Risks/headwinds
• *(Processes, programmes, products, services):* ▪ *Coverage* ▪ *Adoption* ▪ *Completion rates* ▪ *Conversions* ▪ *End-to-end completion* ▪ *Customer satisfaction* ▪ *Diversity Equity and Inclusion Index* ▪ *Quality Index*	• *Business stakeholder resistance requires more insights regularly to buy in fully and bust myths due to commonly held (and dated) beliefs.* • *Inability to provide insights at the speed of business.* • *Security issues: data breaches due to a lack of server policies on data sharing.* • *Slow data validation processes; can't deliver to plan.* • *International:* ▪ *Government regulation (GDPR) requirements.* ▪ *Translation costs.* ▪ *Middle East translation issues; keyboard accessibility/UX issues.*
Template adapted from Bryar, *Working Backwards*[102]	

---

102   Colin Bryar, *Working Backwards: Insights, Stories, and Secrets from Inside Amazon* (St. Martin's Press 2021).

Now you try:

DYNAMIC LEARNING

Look at your goals and the team's goals. What are your challenges in the short term? Are those challenges across the team, and, if so, how might they become resolved with an updated team goal?

Vision	Landscape
Work backwards plan: living into the vision Year 2: Year 1:	
Success measures	**Risks/headwinds**

WORKSHEET 2
REFLECTION: RECALL YOUR PAST DATA PROJECT EXPERIENCES

Use this example and template as a reflective aid to think through past experiences with data. It can be helpful to think on this independently or bring people together as a team to discuss.

Past project:	
What went well?	Example: *We invested in discovery sessions to learn more about our stakeholders' needs.*
What didn't go well? How did it feel?	Example: *Since we operate in silos, we have a lot of redundant projects and resources, making it hard to drive priorities between teams. It's been frustrating, to say the least. Frustration with remaining siloed creates a sense of limited progress, feeling stuck. We could work better if we acknowledge dependencies better and collaborate on joint goals.*
What would you do differently?	Example: *Our ops team was already overwhelmed. We should have collectively taken a step back and tried to align our strategies between groups to highlight mutual needs that impacted our budget and resource requests.*
How do you wish your data partner had consulted you differently?	Example: *Operations can see many redundant requests from our and other teams. I wish we had taken that opportunity to pause, step back and think about everyday needs across the teams to prioritise our requirements.*

WORKSHEET 3
TO-DO LIST BASED NEEDS ASSESSMENT

1.	**What is the specific task, process or deliverable you are hoping to produce?**	List specific tasks or actions. *Facilitate agreement on a standard report template between 5 business units.* *Facilitate agreement on metadata for measures (in particular, definitions and data sources).* *Confirm metadata in a single source of truth; develop an updating process for ongoing change management.*
2.	**Why is it important to meet your departmental or functional area goals?**	List annual departmental goals addressed through this project. *Increase reporting efficiency by 20% in FYxx.*
3.	**Why is it important to meet your annual organisational goals?**	List the annual organisational goals addressed through this project. *Develop a centre of excellence …*
4.	**Why is it important to meet your multiyear strategic priorities?**	List the multiyear strategic priorities addressed through this project. *Develop a centre of excellence …*

WORKSHEET 4
PROJECT VETTING CHECKLIST

Use this checklist to vet your scope against the four project criteria.

Criterion 1: The scope is clearly defined

☐ You can clearly state the objectives and key deliverables of the project.

☐ You have a clear sense of the total number of hours and duration of work (in weeks or months) required to complete the project successfully.

☐ You can state the desired short- and long-term outcomes of the project.

☐ You can clearly communicate the connection between this project and departmental and organisational goals.

Criterion 2: The project meets an important and not (unreasonably) urgent need

☐ You can clearly state the tie between your project's mission and the strategic plan.

☐ The objective or deliverables of this project are important to the data team's ability to demonstrate progress. Without collaborative sponsors, you would not find alternative means to complete the work if cross-functional collaboration were not a possibility.

Criterion 3: The level of field-specific education needed by data team members does not outweigh the benefits of their work

☐ For high-impact projects, the data team and stakeholder group may require increased investment to be successful and adequately prepared to complete the work. Still, a high-impact project result justifies the time spent on training. For example, sponsors and working teams might engage in change management or leadership training for significant change transformation initiatives together. As a team, project management teams might prepare by attending a training in a particular methodology (i.e. 6Sigma, Lean, Agile, etc.) to share a common vocabulary and approach.

☐ For medium-impact projects, the project requires moderate investment (or less) of training and management time for the data team and/or stakeholder group to succeed. For example, the project lead or lead data scientist or engineer might receive training and bring knowledge back to the team.

☐ For low-impact projects, the project requires minimal training and management time investment for the data team and/or stakeholder group to succeed.

Criterion 4: The required members of your team are receptive and available

☐ You have the required skills on the team to effectively manage the data project.

☐ The staff qualified to serve as leads, subject matter experts or decision-makers are receptive and available to help in their respective roles.

☐ One or more sponsors are open to being involved in scoping, securing and guiding the data project.

WORKSHEET 5
SCOPING TEMPLATE

Project name		Owner	
Status			

Scoping statement:

Problem statement (inc. business impact) **Summary problem statement:** **Business impact:**	**Stakeholders** • End customer(s)/Personas • Executive sponsor(s) • Team members
Core success metrics:	**Deliverables/Dates**

		Scope boundaries	
Risks	**Mitigations:**	**IN**	**OUT**

WORKSHEET 6
CHANGE MANAGEMENT PREPARATION TASK LIST

Preparation task	N/A	Completed	In progress	Not started
Alignment with organisational vision defined				
Project outcomes clearly defined				
Business process designed				
Technology or systems designed				
Job roles defined				
Job descriptions written or modified				
Impacted groups identified				
Impacted systems and tools identified				
Implementation start/end dates set				
Measurable objectives defined				
Success/health measures defined				

Since this is a book about data management, the following high-level task list might be appended:

Data sources in scope and identified				
Data acquisition and instrumentation guidelines				
Standards (for each phase of data supply chain)				
Data processing needs determined				
Data analysis steps identified				
Metadata identified and documented				
Long-term storage and backups identified				
Preservation, archiving and disposal datasets identified				
Data sharing, access and release processes identified and documented				
Persistent identifier defined, confirmed				
Monitoring and operations processes identified				

WORKSHEET 7
ASSESSING YOUR ORGANISATION FOR CHANGE

Use the assessment below to evaluate the health of an initiative at the beginning, middle and end of a change initiative to identify key risks. Through repeated application over the lifecycle of the initiative, the PCT Model and assessment enable teams to evaluate progress and improve initiative health with informed interventions.

Rank each factor on a 1–3 scale.

1 = Inadequate 2 = Adequate 3 = Exceptional

Once you have completed Worksheet 7 you should use it in conjunction with Worksheet 9 to determine the risk profile of your organisation. Use Worksheet 8 to think through intended actions.

Likewise, it is important to observe and/or assess the working group as appropriate.

Leadership-sponsorship factors and current scores	1/2/3
1. The change has a Primary Sponsor.	
2. The Primary Sponsor has the necessary authority over the people, processes and systems to authorise and fund the change.	
3. The Primary Sponsor is willing and able to build a sponsorship coalition for the change and manage resistance from other managers.	
4. The Primary Sponsor will actively and visibly participate with the project team throughout the entire project.	
5. The Primary Sponsor will resolve issues and make decisions relating to the project schedule, scope and resources.	
6. The Primary Sponsor can build awareness of the need for the change (why the change is happening) directly with group members.	
7. The organisation has a clearly defined vision and strategy.	
8. This change is aligned with the strategy and vision for the organisation.	
9. Priorities have been set and communicated regarding this change and other competing initiatives.	
10. The Primary Sponsor will visibly reinforce the change and celebrate successes with the team and the organisation.	
Score: (total possible is 30)	

Project management scores	1/2/3
1. The change is clearly defined including what the change will look like and who is impacted by the change.	
2. The project has a clearly defined scope.	
3. The project has specific objectives that define success.	
4. Project milestones have been identified and a project schedule has been created.	
5. A project manager has been assigned to manage the project resources and tasks.	
6. A work breakdown structure has been completed and deliverables have been identified.	
7. Resources for the project team have been identified and acquired based on the work breakdown structure.	
8. Periodic meetings are scheduled with the project team to track progress and resolve issues.	
9. The Primary Sponsor is readily available to work on issues that impact dates, scope or resources.	
10. The project plan has been integrated with the change management plan.	
Score: (total possible is 30)	

Change management scores	1/2/3
1. A structured change management approach is being applied to the project.	
2. An assessment of the change and its impact on the organisation has been completed.	
3. An assessment of the organisation's readiness for change has been completed.	
4. Anticipated areas of resistance have been identified and special tactics have been developed.	
5. A change management strategy including the necessary sponsorship model and change management team model has been created.	
6. Change management team members have been identified and trained.	
7. An assessment of the strength of the sponsorship coalition has been conducted.	
8. Change management plans including communications, sponsorship, coaching, training and resistance management plans created.	
9. Feedback processes established to gather information from group members to determine how effectively the change is being adopted.	
10. Resistance to change is managed effectively and change successes are celebrated, both in private and in public.	
Score: (total possible is 30)	

WORKSHEET 8
INTEGRATION OF PROJECT AND CHANGE MANAGEMENT ACTIVITIES

The list below is a suggested integration of project and change management activities by phase. Use in conjunction with the project method in Phase Three: Manage the Work and Phase Four: Tune the Change. As with all tools in this guide, adapt to your needs.

CM	Start-up phase
	Define project objectives and scope with project sponsor
	Determine high-level timeline and key milestone
*	**Assess the size and nature of the change**
*	**Assess the organisation affected by the change and conduct group member readiness assessments**
*	**Develop a change management strategy**
	Prepare initial project budget
	Select project team
*	**Acquire change management resources**
*	**Assess team competencies in change management**
	Select outside consultants if required
	Conduct team building and train project team
*	**Train the change management team**
	Identify senior stakeholders and members of project steering committee
*	**Identify necessary project sponsors**
*	**Assess sponsor positions and competencies**
*	**Develop the sponsor model and prepare sponsors to manage the change**
	Define communication protocols withing project team

CM	Planning phase
	Prepare draft of project plan
*	**Develop communications plan**
*	**Develop coaching plan**
*	**Develop resistance management plan**
	Review project plan and change management plans with Primary Sponsor
	Finalise project plan
*	**Implement communications plan**
*	**Implement sponsor roadmap**
	Schedule first steering committee meeting
	Prepare executive-level presentation of project plan to steering committee
	Review presentation with primary project sponsor
	Present project plan to steering committee for approval

CM	Research and data-gathering phase
*	**Implement communications plan for this phase**
*	**Implement sponsor roadmap for this phase**
	Define research and data gathering requirements
	Assign research responsibilities to project team
	Conduct current state as-is process analysis (high-level only)
	Conduct member focus groups and manager interviews
	Conduct customer or end-user focus groups and surveys
	Benchmark competitors and similar non-competitors
	Prepare research and data gathering reports
	Prepare executive-level presentation for steering committee
	Review presentation with primary project sponsor
	Present research and data-gathering reports to steering committee

CM	Solution design phase
*	**Implement communications plan for this phase**
*	**Implement sponsor roadmap for this phase**
	Conduct a detailed review of all research findings with project team
	Define the guiding principles and concepts for the future state
	Develop the future state to-be business processes and workflows
	Create a functional needs document for systems and technology
	Create technology requirements document for IT or use in RFPs (request for proposals)
	Define required organisation changes
	Define new job roles and responsibilities
	Prepare a draft of the solution design document
	Review the solution design with the primary sponsor
	Finalise solution design document
	Prepare an executive-level presentation for the steering committee
	Present the solution to the steering committee

CM	Business case and gap analysis phase
*	**Implement communications plan for this phase**
*	**Implement sponsor roadmap for this phase**
	Conduct a gap analysis between as-is processes and to-be processes
	Determine estimated cost savings and revenue growth from new solution
	Acquire bids and develop estimates for systems and technology
	Estimate all implementation costs for project deployment
	Prepare a business case for the new design
	Compare business case results with initial project objectives to ensure alignment
	Review the financial calculations with accounting or finance groups
	Review the business case results and financial assumptions with the primary sponsor
	Send the business case document to all steering committee members for review
	Prepare an executive-level presentation of the business case
	Present the final business case to the steering committee for funding approval

CM	Solution development phase
*	**Complete all process designs and workflows**
*	**Implement coaching plans**
	Develop or purchase required systems and technology
	Conduct trials of each solution component
	Collect feedback from trial users and integrate into the design
	Create detailed job descriptions
	Review job descriptions with HR and legal departments
	Develop a transition plan for the new solution
*	**Define training requirements including change management**
	Develop training curriculum and courseware
	Develop job aids for group members
	Define facility requirements and changes
	Prepare executive-level presentation of transition plan
	Present transition plan to steering committee for approval

CM	Solution implementation phase
	Develop issue tracking and control process
	Train group members on new processes, systems and tools
	Begin phased implementation of solution
	Collect group member and manager feedback on solution
*	**Gather group member feedback on the change process**
*	**Audit compliance with new processes, systems and roles**
*	**Analyse change management effectiveness**
*	**Identify root causes and pockets of resistance**
*	**Develop corrective action plans**
*	**Enable sponsors and coaches to manage resistance**
*	**Implement corrective plan**
	Modify solution design and transition plans based on feedback
	Track and resolve issues during implementation
	Prepare implementation progress report for steering committee
	Measure process and system performance
*	**Celebrate early successes**
	Measure business outcomes and compare with project objectives
	Adjust the implementation to achieve desired business results
	Prepare executive-level presentation of process, systems and business results
	Present final business results to steering committee
*	**Conduct after-action reviews**
	Begin phased process of dissolving the project team
	Turn over control of new processes and systems to appropriate line managers

WORKSHEET 9
PCT ASSESSMENT SUMMARY AND RISK PROFILE

In conjuction with Worksheet 7: Assessing your organisation for change, summarise your findings, determine your risk profile and think through intended actions.

Summary:

Leadership score	
Project management score	
Change management score	

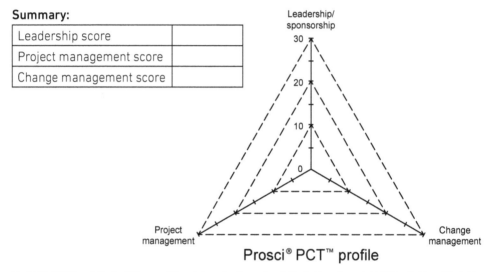

Prosci® PCT™ profile

ProSci®, PCT™ and 'Project Change Triangle' are trademarks of ProSci Inc. Copyright © ProSci Inc. All rights reserved.

SCORE

10–19 High risk – Jeopardy – Immediate action needed
20–21 Alert – Possible risk – Needs further investigation
25–30 Strength – Leverage and maintain

RISK PROFILE AND INTENDED ACTIONS

List factors that score a one, then tally.

Element (L/S, PM, CM)	Factor #	Intended actions

WORKSHEET 10
PCT SPONSOR ASSESSMENT

Use this assessment to determine the current sponsorship competency of each sponsor. If there is not enough information to complete the worksheet for a sponsor, request a meeting with the sponsor. Use these questions as a discussion tool and ask them to assess their behaviours on current or past projects.

Participated actively and visibly throughout the project	Least			Most	
	1	2	3	4	5
Attended project and status meetings regularly.					
Was present at kick-off special events and training sessions.					
Held the team accountable for results (defined objectives, conducted ongoing reviews).					
Was involved in critical decision-making (at critical milestones, at steering committee meetings, in 1:1 sessions).					
Ensured the project had the right team members, budget and resources for success.					
Was accessible to the project team; ensured that other managers were accessible as well.					

Built a coalition of sponsorship with peers and managers	Least			Most	
	1	2	3	4	5
Sponsored the change with direct reports (created awareness of the need for change, built support and followed up).					
Established clear expectations with mid-level managers.					
Dealt with problem managers (managed resistance).					
Created a sponsorship cascade with managers, ensured that they were building support with their direct reports.					
Listened to and addressed management concerns.					
Held direct, face-to-face meetings with managers to explain 'what, why and how'.					

Provided management leadership team with frequent updates and status information.					
Sponsored the change upward.					

Communicated directly with data teams and business stakeholders	Least			Most	
	1	2	3	4	5
Effectively communicated why the change was happening, the risks of not changing and the vision for the organisation.					
Linked key performance indicators and financial objectives to the change.					
Enabled communications to be two-way (allowed for feedback and question/answer sessions).					
Spoke face-to-face at town meetings, road shows and key presentations.					
Communicated frequently throughout the project and with multiple media (not just during project kick-off).					
Interacted effectively with managers; helped them create and communicate a consistent message to group members, extended stakeholders and end users.					

Use the overall score of each assessment to assign a competency level to each sponsor using these scores:

- **Level 1**: Score of 80 or higher – Sponsor is **knowledgeable and experienced** in change sponsorship.

- **Level 2**: 70–79 – Sponsor has **limited knowledge and experience** in change sponsorship.

- **Level 3**: Score of 69 or lower – Sponsor has little to **no knowledge and experience** in change sponsorship.

Low scores indicate what upskilling efforts the project lead must support. The project lead will prepare communications, job aids and training to enable sponsors behind the scenes.

NOTE: Both the diagram and assessment should be used very carefully as they could embarrass or surprise key business leaders.

WORKSHEET 11
PCT GROUP MEMBER ASSESSMENT

Use the assessment to determine the current competency of each group member. If there is not enough information to complete the worksheet for a group member, use as a discussion tool and ask them to assess their behaviours on the current project or past projects.

Awareness		Strongly disagree			Strongly agree	
	NA	1	2	3	4	5
I understand the business reasons for the change.						
I understand the risks of not changing.						
I understand the impact on my day-to-day work activities.						
I understand why this change is happening.						
I understand why this change is happening right now.						
I understand the triggers (internal and external) for this change.						
I understand our leader's vision for this change.						
I understand how this change aligns with our strategy.						
I understand the intended results of implementing this change.						
I could explain why this change is happening to a colleague.						
The person who has explained the need for change is credible.						
All the questions I have about why the change is happening have been answered.						
I understand the nature of the change's impact on me, my work and my team.						
Communication about why the change is happening has been effective.						

Desire		Strongly disagree			Strongly agree	
	NA	1	2	3	4	5
I am personally motivated to be part of the change.						
I look forward to the new, changed environment.						
I believe in the case for this change.						
I am supportive of this change.						
I know What's In It For Me (WIIFM).						
I have made the decision to participate in this change.						
I believe that if we take on this change, there is a high likelihood that we will be successful.						
I understand what is in it for me, for my team and for our organisation.						
I have conviction in the reasons for change.						
I believe my teammates/peers support this change.						
I believe my manager supports this change.						
I believe my senior leaders support this change.						
I believe the organisational motivations for this change are true and accurate.						
I am personally motivated to be part of this change.						

Knowledge		Strongly disagree			Strongly agree	
	NA	1	2	3	4	5
I have the skills and knowledge to be successful during the change.						
I have the skills and knowledge to be successful after the change.						
Training has been adequate to prepare me.						
I clearly understand the impact this change will have on my behaviours, processes, tools and workflow.						
I have the knowledge I need to be successful while the change is implemented.						
I have the knowledge I need to be successful after the change is implemented.						
I do not foresee any knowledge gaps that might make me less successful because of this change.						
I have received adequate training to feel prepared to be successful.						
I have the foundational knowledge I need to be successful.						
I have the project-specific knowledge I need to be successful.						
Resources and tools are available to help me be successful.						
I know where to go for additional knowledge and support.						
I have the capacity to learn the new things I need to be successful in this change.						

Ability	NA	Strongly disagree			Strongly agree	
		1	2	3	4	5
I can perform the new duties required by the change.						
I can get support when I have problems and questions.						
I have practice at performing in the new environment.						
I believe that I can close the knowledge–ability gap.						
I have been able to practice new skills and behaviours.						
I am able to make the necessary changes to my behaviours, processes, tools and workflows.						
I believe the training provided will give me what I need to be successful in this change.						
I can access additional knowledge and support when needed.						
I can access coaching on the change from my manager when needed.						
I can overcome any barriers to implementing the change in how I do my job.						
I have the capacity to implement the change in how I do my job.						
I know where to go for help if I have problems making the change.						

Reinforcement		Strongly disagree			Strongly agree	
	NA	1	2	3	4	5
The organisation is committed to keeping the change in place.						
I know the consequences of not performing my new activities.						
I am rewarded for performing in the new way.						
There are mechanisms in place to sustain the change.						
My performance in the new way is evaluated.						
I have an outlet for providing feedback on the change.						
The new behaviours, processes, tools and workflows are in my performance review system.						
I am committed to sustaining the change.						
I believe my teammates are committed to sustaining the change.						
I believe my management is committed to sustaining the change.						
I believe my senior leaders are committed to sustaining the change.						
I feel the incentives that may be opposed to the change have been addressed.						

NOTE: Both the diagram and the assessment should be used very carefully, as they could embarrass or surprise key business leaders.

WORKSHEET 12
MAPPING DISCOVERY AT THE END OF PHASE THREE

What are the three primary questions to answer through discovery?	Who will have the appropriate information to answer this question?	Internal outreach list and owners
Key Question 1	Name, Group Name	
	1	
	2	
	3	
	4	
	5	
Key Question 2		
	1	
	2	
	3	
	4	
	5	
Key Question 3		
	1	
	2	
	3	
	4	
	5	

WORKSHEET 13
MISSION AND GUIDING PRINCIPLES (OBSERVE)

Summarise your collaborative effort determining a mission and choosing the principles that will guide your behaviour and decisions as you embark on your data projects.

Developing a mission statement. Point your compass in the right direction as early as possible. As a team, decide the 1–3 things your strategy needs to accomplish and succinctly summarise it/them into one sentence.

The _____ mission is to

(initiative's or team name)

bring data to the _____

(your target)

_____, and

(goal 1)

_____.

(goal 2)

Determining guiding principles. As a team, choose 3–5 guiding principles from your organisation or inspired by outside organisations. Keep track of how they work and don't work throughout the process. By observing what does and does not work, you are starting the process of right-fitting what data-forward looks like for your organisation and making it easier for others to adopt.

Proposed guiding principles	What works	What does not work

WORKSHEET 14
THREE HOWS – MULTIYEAR GOALS (PRACTICE)

Review Table 6.7 that illustrates how this process plays out a data-as-a-service requirements list for your organisation, and adapt it for your needs. How does your data project map to team, functional and organisational goals?

What is the multiyear goal or strategic priority?	
	(by)
How will you address this annual organisational goal in the next year?	
	(by)
How will the annual organisational goal be addressed through annual department goals?	
	(by)
How will you meet department goals through specific tasks, activities or needs?	

WORKSHEET 15
TO-DOS LIST (PRACTICE)

The table below demonstrates how to consider possible sources of data requirements to begin generating a list of high-priority opportunities. From there, you can develop a way to rank their urgency.

Below is an example template. Modify to your needs.

List all potential projects or data initiatives	Tactical (bottom-up)	Strategic (top-down)	Urgent	Important	Risk level
1					
2					
3					
4					
5					

WORKSHEET 16
AUGMENT CORE COMPETENCIES (LEARN)

Review Table 6.15 to recall how successful data projects require combined effectiveness in three areas: business strategy and stewardship, project management in data functions and change management impacting culture and the pace of change. Use the table below to think through your needs – asking for help requires forethought.

Core competency	Use one or two sentences you would describe to a sponsor	Given this core competency, how will the need be scaled and how can the sponsor help you build the right path?	Given this potential scaling need, what specific characteristics would you look for in a stakeholder?
Example: *Learning and development project manager or curriculum lead.*	*Need an enablement partner to review and consult on training offerings for the current cross-functional stakeholder group.*	*Need a learning and development partner to help convert a current training programme into a workshop format to be shared across the organisation instead of just our stakeholder group. We need a partner to help us standardise and scale our learning materials to be included in org-wide onboarding training.*	*A good fit would be a training resource from: the product management training group, human resources/ employee experience lead or user experience lead.*
1			
2			
3			
4			
5			

WORKSHEET 17
DATA SUPPLY CHAIN: CORE SYSTEMS

The table below demonstrates how to consider common areas of focus in developing support systems.

Repeatable systems lay the groundwork for ongoing efficiency, learning and scale. Determine what systems and processes you would like to try and think about why you want to try them. Consider their key design factors and the value they will contribute. This is the time to standardise the process behind all data efforts.

Core system	Why is this important?	Key design factors

INDEX

Page numbers in italics refer to figures or tables.

www.ingramcontent.com/pod-product-compliance
Lightning Source LLC
Chambersburg PA
CBHW041007050326
40690CB00031B/5291